100
TOP CROPS
YOU CAN GROW

Mark Valencia
with Kate Di Prima

"The moral rights of the author have been asserted in accord-
ance with the Copyright, Designs and Patents Act 1988 in
Australia, and the Copyright Act 1994 in New Zealand".

9781761280023 (print)
Designed and produced by Brio Books (Booktopia Direct)
34-38 Cosgrove Rd Strathfield South, NSW

Cataloguing-in-publication data is available from the
National Library of Australia

Printed in China

Contents

The story behind self-sufficient me

Some of you will already know me, but for those who don't, I'm Mark Valencia, a retired soldier turned self-sufficient home gardening YouTuber with over 2 million followers around the world! Yes, I pinch myself every day and feel honoured that so many people are interested in my self-sufficient journey and the information I share online.

Creating a beautiful garden at home is one thing, but when your garden becomes the provider for you and your family, well, that's another thing altogether. People often ask me, 'Why gardening and what inspired you?'

Well, you can't ignore the health aspects of getting outside and growing organic food. But to be brutally honest, in the beginning, my main goal was just to save money on our grocery bills. It'd be easy to romanticise the situation and say that it was always the plan to buy a plot of land, leave my military career, grow a tonne of food in our backyard and become a successful YouTuber.

But that wasn't the case at all.

Back when I started, pure economic survival – and the desire to give our children a more stable and healthier lifestyle – was the focus for my wife, Nina, and me. As you can imagine, family and friends were a bit worried – some thought we were downright crazy – but we never doubted for a second our decision to transition to what we were sure would be a better life for us all.

I was born in Toowoomba in Queensland after my biological Croatian father "did a runner" and disappeared from the opal fields of Lightning Ridge, NSW leaving my mother, Coral, to raise me on her own with the help of my grandparents Herbert and Isabel Bruggemann. When I was 4 years old, we moved to Darwin in the Northern Territory with my stepfather (Robert) who was a Chilean immigrant, hence my last name Valencia. In Darwin, I became addicted to salty plums and all sorts of oriental delicacies thanks in part to the heavy Asian influence up that way. Back then, Darwin was an amazing mixing pot of people from all over the world and no doubt it still is. The closest family we had up there at the time were my Italian uncle Guido Zuccoli, his wife, aunt Lynette, and their daughter, cousin Annamaria. The Zuccoli family were well-known Northern Territorians with my uncle being a prominent businessman and aerobatic pilot and my aunt an accomplished artist. With regular gatherings and backyard BBQs, I had my fair share of Mediterranean food influences too, living off olives, salami, ravioli and antipasti. I never recall going hungry!

In 1974, when I was 6, Cyclone Tracy hit Darwin on Christmas Day. I remember the night vividly as my parents, baby sister (Ines) and I moved from room to room trying to find a safe place to bunker down. The storm devasted the area and wrecked our government rental home, rendering it unliveable. Tragically, sixty-six people lost their lives. We were lucky to survive and if it wasn't for two power poles falling either side of our home with the wires between them securing the roof that was beginning to lift, I might not be here today. Shortly after the cyclone hit, my uncle Guido piloted a plane and flew us out of devastated Darwin to Toowoomba where we stayed with my grandparents. Several months later,

we returned to Darwin to re-build, my younger brother, Philip, was born and then when I was 9, we moved back to Toowoomba to live again with my grandparents.

I subsequently spent a lot of time with my grandfather who was a retired farmer, watching and helping as he grew fruit and veggies on his small urban block. I also visited my uncle Ernie's dairy farm during school breaks. I used to ride my bike around 30 kilometres from home to their farm at Groomsville near Crows Nest, QLD to milk the cows at 3 a.m. My friends thought I was bananas! But even as a kid, I was fascinated by how food was produced, and subconsciously I think I understood the philosophy or importance behind self-sufficiency.

In my early teens, my mother and stepfather got divorced leaving my mum to raise us three kids with little capacity or funds to move out from our grandparents' place. At the time, I did understand that we must have been a bit of a burden but all I ever felt was love and affection from my grandparents, aunties, uncles and of course, our mother.

One day, my mother called me into my grandparents' bedroom (I thought I was in trouble). She said she had something important to tell me. And that was – we won the lotto! Yeah, I know, amazing hey . . . It wasn't a lot of money, but mum won enough to buy a house (fully furnished) and a brand-new car. Just like that, we were suddenly independent living in our own home. God has strange ways of answering prayers.

My teenage years were memorable, not least for our extended family get-togethers for special occasions. These events were always centred around food with an emphasis on quality produce and making dishes by hand. Back in those days takeaway was a rare thing, going out for dinner was even rarer, and it was a given that my mum (a very capable cook) would *always* make breakfast, lunch and dinner for us kids. I never felt like I missed out on anything and still feel like my childhood was extraordinarily happy. Academically, I was a poor student (to put it mildly) I struggled to take learning seriously, but I made lifelong friends that are still extremely close to me to this day. Some of my best early memories growing up in my hometown of Toowoomba are about spending time with family, friends and eating lots of good local food.

In 1987 at 17, I joined the Australian Army because soldiering was about the only thing I thought I would be good at. I was enticed by the outdoor lifestyle, fitness, and limited educational requirements. In 1992, I was posted to Darwin where, as fate would have it, I lived with my Italian uncle Guido, who kindly offered me free boarding at their luxury townhouse near the city in Stuart Park – I was the envy of all my Army mates. Guido was winding up his successful construction company at the time and planning to move to Toowoomba to semi-retire with his family. During those few years, my uncle influenced my food journey greatly. Guido's love of local fresh produce and home cooking was infectious! He taught me several Italian recipes and his tips, food ideas, chats about life, intelligence and overall gentlemanly nature will stay with me forever. Guido sadly passed away on March 6th, 1997, when his newly restored T6 Harvard crashed soon after take-off from Tindal Air Force base.

Over the next decade or so the Army took me to many remote and interesting places. These experiences couldn't help but shape me for good and bad – but mostly good. I encountered a lot of different cultures at close range, something that not many people get to do, and I got to appreciate how others live and survive. I learnt about their traditions, their beliefs and of course their food.

In May 2000, I was involved in a freak parachuting accident during a training exercise over central New South Wales. My right arm was nearly severed when exiting the door of the plane due to a loose static line becoming tangled around my wrist as I jumped out. It goes without saying landing on the ground with one arm wasn't what you'd call textbook, and I did more damage as I hit the airstrip like a sack of spuds. It was a long and difficult road back from that accident but with the help of some amazing doctors and nurses, Nina, my mum, sister, brother, extended family and caring friends I somehow recovered sufficiently to continue serving in the Army.

I wouldn't say that I suddenly developed a better outlook on life after the accident – I've always been thankful for what I have – but in hindsight this event and the accumulation of experiences up to that point sharpened my focus on what was truly important. So even though I served another eight years in the Army and ended up being promoted, I'd already begun to plan my exit and the next phase of my life.

When I finally did leave the Army, the idea wasn't necessarily to find a place and grow food. Originally, Nina and I just wanted to slow down a bit after clocking up a combined 38 years in the military – and settle somewhere less 'busy' and more peaceful. We also wanted a more stable, stationary lifestyle where our kids were at home and not at childcare 24/7. So we bought a three-acre property about an hour's drive north of Brisbane and began a brand-new life away from the stress of the city.

There were definitely challenges in the early days. I took a considerable pay drop for one and we became a single-income family with two kids. At the time, Nina was working full-time as an army reservist *and* a Red Cross nurse. I felt the best thing I could do – besides taking on the lion's share of household duties such as cooking, cleaning and getting the boys to school or sport – was to step up and provide for the family in other ways. We quickly worked out that food was one of our biggest expenses.

I thought to myself, 'Why not utilise the space we have to grow as much food as possible?' So I started building raised garden beds and planting as many fruit trees as I could. Interestingly, I discovered by accident that gardening also helped me relax. It really calmed me and helped with the PTSD which I'd acquired from several incidences and my accident during my time in the military. This added benefit was a surprise, but the bottom line was exactly that . . . the bottom line! Once I realised we could grow so much

produce from cheap packets of seeds, I knew that we'd be making some massive savings.

From there, other benefits followed: the positive mental health side of things; the physical exercise; the vitamin D (sunlight); the plentiful organic and pesticide-free food and a beautiful garden to be proud of. Possibly the best thing about growing your own fruit and veggies if you have children is getting more diversity into their diet. Eating food grown in your own backyard is a powerful and positive message. Because kids see, they pick and they eat. If they're inquisitive – and most kids are – they get excited about food. We don't have any fussy eaters in our family because of that. My boys, who are now 21 and 18, still ask me if the food on the table has come from our garden.

If I had to pick the best thing out of all the incredible benefits that have come out of our backyard garden, it would be the positive influence healthy eating has had on our boys. Growing evidence (pun intended) has shown that chemical nasties in our food – whether that's pesticides, fungicides, additives, micro-plastics or over-processing – have been linked to ADHD, cancer and many other preventable diseases.

Today, I'm certain that the pain of leaving my military career in the hope that we could make healthier lifestyle choices for our family was the right decision – even if it was a tough path to travel at times.

—

Our home garden never gets boring and the effect it has on me is incredible – every day, you're witnessing nature from its very beginnings! I still marvel at how a tiny coriander seed, for instance, can turn into a big, beautiful and edible plant. It's a miraculous and complex process as that dormant seed receives a few essential elements and slowly becomes a fully functioning, living thing.

Another question I'm often asked is, 'Where did the motivation to become self-sufficient come from?'

Well, in modern society it's almost impossible to be self-sufficient in *everything*, so that's why I always say, 'Don't worry too much about that; just be self-sufficient in *something*.' You'd be surprised how being self-sufficient in a few simple things like herbs or veggies can make a big difference.

Growing food ain't rocket science and once people understand this, most get the hang of it pretty quickly. It really doesn't matter who you are or where you live. You could be in a city apartment, an urban block, a semi-rural acreage or on a large farm. Wherever you are, with a bit of time, effort and ingenuity, you can grow your own food! And that's the message I want to spread.

—

'Look and see the earth through her eyes . . .' was the first 'one liner' I made up to explain the value of watching how nature does things and then working with her to solve problems and grow better fruit, vegetables and herbs. The slogan helped us develop our 'eye' logo with the iris being the earth. Some people think the logo is a bit creepy, and a few weirdos have tried to connect me to the Illuminati, but the truth is I settled on our logo simply because I thought it best described our values. And there's no doubt that it's 'eye-catching'!

What's this book about?

In a nutshell, I write about my top 20 food crops in five categories. All in all, I cover 100 crops, most of which you can grow wherever you live. I felt this was an appropriate way to structure the book as some of my most popular YouTube videos are numbered lists – and who doesn't love a countdown? I also cover some of the disciplines you need to understand for big, healthy harvests of quality produce. And I offer a heap of essential tips too. The start of the book lays the groundwork for better gardening based on my practical experience over the past 20 or so years.

There are thousands of food crops you can grow at home – and I've tried more than most – but we have to draw the line somewhere. That's why this carefully considered, narrowed-down list was such an interesting challenge. It was so good, in fact, I thought it'd make a great book!

Dietitian Kate Di Prima has been involved with the project right from the start. It was Kate who urged me to put pen to paper in the first place. Without her persistence and motivation – and her experience in publishing – this project may never have got off the ground. Writing a book not only takes

time, it's also a process that needs to be learnt – so you could say I needed help!

Nina and I have known Kate and her husband, Paul, for over a decade and we've become close friends, mainly because they're both happy and fun people to be around, but also because they share our love of healthy food and a healthy lifestyle. Kate brings 30 or more years of nutritional experience as a practising dietitian to the table, explaining the nutritional value and uses of the food crops I detail in my lists.

Over the past couple of decades, I often thought about writing a book on growing your own food for self-sufficiency, but I just didn't have the time and experience to get started. That all changed a few years ago when Kate rocked up to our place with her voice recorder and proceeded to pepper me with questions beneath the shade of a Lane's Late Navel orange tree.

You could say the rest is history . . .

A note from Kate Di Prima

Some years ago I was lucky enough to meet Mark and Nina Valencia at a BBQ at a mutual friend's house. They walked in carrying huge boxes of fresh fruit, vegetables and salad items. Following behind Mark and Nina were their two boys who were also carrying bags of vegetables. My first thought was, 'Oh how lovely, they've been to the markets.' There were some exchanges and grateful hugs between our friends and Mark and Nina about how wonderful the generous offering was. I thought, 'Gee, they must be *really* good friends as that would have cost them a fortune!'

It wasn't until later in the afternoon that I realised that all the food they'd brought had come from the Valencias' own home garden. Then I was impressed!

Years later and now firm friends, Mark, Nina, Paul and I were fortunate enough to be travelling together in Italy. One day we enjoyed a lunch prepared high in the hills of Positano on the beautiful Amalfi Coast. The lunch involved just about every member of the family running the place, and the ingredients they used were picked that morning from the hill on which they lived. Mark and I talked about the fact that both our families were raised on healthy core foods – many that could be grown at home – and that growing your own food was something *everyone* should try. And so the seed was planted for a book that would help any budding gardener grow at least *something* at home – and then prepare a healthy meal for themselves and their families.

My husband Paul and I are in a very different situation to Mark and Nina. We live in inner-city Brisbane with very little useable garden for growing food. However, Paul, who is Sicilian, has built some above-ground garden beds on our front deck. We also have three potted citrus trees out the back. I'm nurturing them daily and watching these beautiful trees bear fruit.

Some foods are easier than others to grow, but with Mark's guidance we will continue to experiment with producing our own fresh produce in the inner city.

A little bit about me

I was born and bred in Canberra, where my dad was in the public service and my mum was a stay-at-home mum – and a great cook. I used to watch her make lasagne, moussaka and spinach pies while I ate breakfast before school. The timer was set for the oven to come on at 3 p.m. every day, and dinner was ready on the table at 6 p.m. We lived in a little three-bedroom brick house on the outskirts of the city, with a large backyard where Mum and Dad cultivated a beautiful veggie patch and grew some stone fruit trees. Mum made plum jam and crab apple jelly to sell at school fêtes or to friends. We would eat beans straight off the vine and strawberries straight from the ground . . . it was truly delightful.

Many years later I moved to Brisbane for my first job as a registered dietitian. Around that time I also met my husband, Paul, who was a chef in the resort town of Noosa. We both adore food, and a dietitian and a chef is a match made in heaven . . . most of the time! Paul has taught me how to cook and I have taught him about the nutritional benefits of foods. Mark taught us how to set up a garden in a limited space and motivated us to try to produce some of the salad and vegetable items that we love. Having our 2 children soon after we were married chartered me down the Paediatric Dietitian pathway and spurred my desire to help parents feed their fussy kids. I couldn't believe how many children refused to eat fruit, vegetables and salads. A few years later I co-authored my first book *More Peas Please, solutions for feeding fussy eaters*. Like Mark and Nina, we didn't have fussy children when it came to food so helping families overcome these food obstacles has been extremely rewarding. When it came to writing this book, Mark asked me to contribute some ideas about the nutritional benefits of nourishing homegrown produce, and how to use all of the wonderful fresh salad, vegetables and fruits that you can grow.

The phrase 'food is love' is something that has resonated with me all my life. My wish is for everyone – from little ones starting out their journey with food to the elderly members of our community – to gather around a dining table and enjoy a healthy wholesome meal together, especially if you have had a hand in producing it yourself.

Food brings people together on many different levels. It's the nourishment of the soul and body: it's truly love.
Giada De Laurentiis

Planning and siting your food garden

Lots of things in life involve compromise and food gardening is no exception. Rarely is there a perfect urban backyard that's 100 per cent suitable for growing fruit and veggies. But the good news is, growing food on a small scale means we don't have the commercial pressures of a retail farm to worry about and that's a *big* advantage!

You'd be surprised at what you can grow in less-than-ideal conditions. Plants that usually like full sun often do quite well in part-shade. Fruit trees that are native to a particular climate such as tropical or cold may still produce good enough fruit in subtropical or temperate conditions.

So what are the essential things you need to do when planning and siting your veggie garden?

Firstly, never try to 'hide' it. Many people over the years have written to me worried that their veggies aren't growing, and my first question is always, 'Where have you put them?'

I ask this because a surprising number of people tell me they've stuck their veggie garden or small fruit tree orchard at the back of the shed away from the house in a place where it can't be seen. The reason they do this is that vegetable gardens can look messy and informal. That's undeniably true!

I happen to love the look of a messy food garden. I think vegetables and fruit trees are as pretty and interesting as ornamentals. But growing food is not about looks – it's about *productivity*. And that's why sunlight is essential. Siting your garden with an emphasis on how much sunlight it receives rather than hiding it away is the only way to go.

Vegetables and fruit trees are also hungry; apart from the sun, they need nutrients and water in order to grow well and produce healthy crops. But, above all, a vegetable garden or fruit tree orchard must be situated where it'll get the most sunlight. If this doesn't happen, the plants and trees won't receive the energy they need to thrive.

Furthermore, if your food garden is stuck 'out of the way' somewhere, then chances are it won't get the attention it needs because it's human nature to forget to water, fertilise, weed or notice pests until it's too late. Don't get me wrong; vegetable gardens don't have to be high maintenance. But they *do* need regular care if they're to grow as much quality produce as they should.

So the takeaway is if you build a veggie garden, please don't hide it – especially don't hide it in the shade! The ideal approach is not to care where your vegetable garden goes, even if that means it's in the front yard for all to see. I'd put it there if that's the spot with the best sunlight and ease of access.

As I mentioned at the start of this section, backyard food gardeners have flexibility for where they build their vegetable garden or plant their orchard. But what exactly is 'best sunlight' and can we compromise on this? If the sunniest spot on your property happens to be the driveway, obviously that won't work. But if you put your veggie garden where it can get at least six hours of direct sunlight per day, this should be enough. For many homes, the sunniest spot is usually in the middle of the backyard – the space that often competes with the swimming pool, cricket pitch or

clothesline. Siting a vegetable garden in the right spot can be tricky.

Thankfully, and depending on the sun, most urban Australian blocks have enough room to compromise and at least squeeze in a few raised garden beds. So before setting your food garden in stone, watch the sun to see how it moves across your property (preferably during the winter months when the sun is lower in the sky) and choose a spot you know gets enough sunlight. Remember: the more sun, the better. If you can position your veggie garden in a spot that gets eight to 10 hours per day, that's fantastic, but as I say, six hours is the minimum. You *can* grow a veggie garden on a balcony with less than that, but you need to be more selective with what you plant. Certain salad crops will do okay in a shaded area, and some fruits, such as passionfruit, can grow in part-shade. Pineapples also grow well in part-shade, they just take a bit longer to fruit.

Another important tip is to make sure you have good access to your veggie garden. There's no point having it in full sun if it's 200 metres from your back door, requiring you to take a cut lunch and a water bottle every time you go visit! Lack of 'presence' in the veggie patch can be why animals like birds and possums, for example, wander in and eat everything in sight. There's no one walking nearby, no scents, family pets or noise, so the animals come in and decimate everything.

If your veggie patch is too far away, you also might not be bothered walking all that way to tend to it. We've got a big veggie garden and a decent watering system, but I only use it if we're going away, and then I'll stick it on a timer so it comes on when needed. Most of the time, I prefer to hand-water our crops. That way, I get to walk around my garden, connect with nature, and feel the heartbeat of the plants. It's a type of meditation for me, but it's also a way to examine each plant or garden bed and get a feel for how they're going. I also 'spot water', so I don't waste it. In other words, I water when a plant needs it and hold off if it doesn't.

Being present in your garden gives you an edge. It allows you to notice the subtle changes so you know what needs watering, you can spot a disease or pest infestation before it gets out of control and work out when your plants need fertiliser or other minerals and nutrients.

I'm not a huge fan of 'permaculture rules' because I find some of them irrelevant and unattainable for the average food gardener. But one rule I do like is to grow something you

harvest regularly closest to the house for quick and easy access. Other veggies that you don't need on a daily basis can be sited further away. For example, position your herb garden close to the back door so you can easily grab whatever you need on a daily basis. Put your corn, carrots, potatoes and other seasonal produce a bit further away.

Without meaning to contradict myself, I do like to add some herbs to my central garden too because they attract beneficial insects, such as bees, that help to pollinate the veggies. Herbs also work to repel or 'hide' your plants from any harmful, destructive insects. Like with permaculture, I'm not a huge believer in 'companion planting' – the idea that certain crops planted together develop a symbiotic and beneficial relationship. But I do think that some food crops or plants like herbs and flowers can repel pests and encourage good animals to visit. Predator insects, grub-eating birds, spiders, lizards, frogs, worms and even some pets can play a positive role in protecting and creating a thriving garden. So finding ways to attract them is a good idea.

To sum things up, here are my top three takeaways for where to put your veggie garden:

1. **Don't hide it** – make your food garden a feature of your home! I didn't mention it before but always start by growing things you and your family actually like. There's no point planting beetroot if you hate the taste!

2. **Ensure your food garden gets plenty of sunlight** – at least six hours a day.

3. **Make your food garden accessible** so that you (and helpful animals) regularly visit your plants, and don't forget they're there.

Growing method

Once you've decided where to put your garden, the next step is to choose your growing method. Most people grow crops in soil either inground or in containers, but there are other ways like hydroponics and aquaponics that are a bit beyond the scope of this book. Check them out online if you're interested. For the beginner gardener, I recommend starting your growing journey by planting your fruits and veggies in soil or potting mix. Less can go wrong, and you'll acquire a basic understanding of gardening principles before moving onto more complex growing methods.

In case you didn't already know, I'm a huge raised garden bed fan. Growing food in raised garden beds solves many problems right off the bat. If you have the space and are reasonably fit and healthy, of course you can plant straight into the ground like I did when I started our vegetable garden back in 2006. Just dig up a patch of grass and prepare the patch just like the farmers of yesteryears when such things like raised beds weren't available. But if you're in a rental property or only have a paved or concrete area – or if bending over and pulling out weeds and grass runners annoys you – I'd suggest raised garden beds. A big deterrent for people growing their own fruit and veggies is that gardening becomes difficult and labour-intensive. Having a garden bed that is waist high makes sowing seeds, planting, pruning, weeding and harvesting much more accessible compared with bending over or gardening on your knees, especially when you're getting on a bit or you've got leg or back problems.

Gardening in raised beds has exploded in recent years. Why? Because it works so well! I was one of the first YouTubers to start making regular videos on the many benefits of high-sided raised bed gardening. One of the main reasons more people don't do it is the cost. Depending on the materials, the price of a raised bed can be high compared to planting straight into the ground – which is virtually free. Having a car is expensive to buy and maintain while walking everywhere is free, but it's like the old saying, 'Why do it hard when you can do it easy?' I would argue that the extra cost of raised beds is worth it if it makes gardening less painful, easier, more fun and more productive.

There are a tonne of different materials and DIY kits on the market these days to make raised beds using wood, plastic, concrete, Aluzinc or steel. All these products have advantages and disadvantages, but my favourite material is Aluzinc, followed by recycled plastic. Aluzinc steel ribbed garden beds dominate my vegetable garden because they're good quality, long-lasting, good-looking, strong but lightweight, and easy to install or move if required.

Whether you're planting directly into the ground, hilled up furrows or in a raised garden bed, you don't need to go very deep for your vegetables. Most will grow in less than 30 centimetres (12 inches) of soil, which is about the length of a kid's school ruler; that's all they need. Exceptions like asparagus need a bit more depth, but even large tomato plants can be grown in shallow pots. Of course, a deeper garden bed up to half a metre (20 inches) of suitable soil/medium is better for a large fruit-size, 6-foot-high tomato plant, but most traditional veggies, particularly salad crops, onions and brassicas need minimal depth.

What soil (growing medium) should you use?

Although most veggies don't require deep soil, the *quality* of what we call the 'growing medium' is really important. Just as a farmer wanting to grow a crop in a big paddock looks for good topsoil, a backyard food grower should look for the same qualities in their inground garden, raised beds, elevated planters or pots. Your topsoil doesn't necessarily have to be full of nutrients because you can always add compost and fertilisers to improve it.

However, it *must* have good nutrient-holding capacity and not be too heavy/dense or too light, especially for vegetables and fruit trees, as their roots need oxygen and the ability to grow freely through the soil. The perfect soil is friable (easily broken into small pieces), loose dirt with a crumbly texture containing enough structure to support a root system. A compact, hard soil like clay that does not crumble and is sticky and heavy when wet will suffocate the roots and starve your plants of nutrients.

Make sure your soil isn't too sandy either, as sandy soil lacks structure and won't hold water or nutrients. Nutrients get 'washed' through, making it harder for plants to find food. Lighter soils also heat up quickly and dry out faster, and can damage young plant roots. Good soil is essential as it encourages fungi, worms and other microbes, and increases the nutrient density of the earth. The best growing medium is almost 'Goldilocks' type soil: not too hard and not too soft, not too wet and not too dry! Just right, in other words.

Over time, you can turn bad soil into good soil by adding compost, animal manures and mulch, or burying organic matter such as kitchen scraps, but this can take a long time – sometimes years. If you don't have good soil initially, buy some premium topsoil from a garden store or landscaping supplies centre. Ask for quality garden soil in bags or have it delivered in bulk. Bulk delivery by the tonne or trailer load is much cheaper than bags. Make sure that the soil you buy is premium and not some inferior home-mixed product of cheap sand and animal manure. Typically, premium garden soil from a quality landscaping supplies centre should contain plenty of compost and organic matter with added minerals like some tiny rocks and be dark black or grey. Moist soil should hold together in a ball if crushed in your fist but then crumble apart when poked with your finger. It shouldn't be sticky or easily fall through your fingers like sand through an hourglass.

If you intend to grow your fruit and veggies in high-sided raised garden beds (typically 70–80 centimetres or around 32 inches), you'll find that these beds need a lot of soil to fill. And since most veggies *don't* need a growing medium that deep, it's a waste to fill the bed using all premium soil. To save money, fill the bottom half or two-thirds with a cheaper soil mix, some rocks or drainage material like crushed granite. Raised beds don't need drainage in the base; gravel or crushed granite would only be for fill purposes. The other way to save money on filling a raised garden bed is to use an old European method called 'Hügelkultur' where logs and sticks and other organic waste are placed in the base of the bed or mound (more about this method later on).

Regarding fertilisers, I think it's a waste of money to buy bulk soil with added fertiliser. While it might save you some time buying it pre-mixed, adding fertiliser yourself when required is easy. You can also choose what you want to add when you need to, instead of accepting soil blended with generic synthetic-based nutrients or an overzealous mix of raw chicken poop that inadvertently burns your plants! I like to use our homemade chicken manure or trusted organic fertiliser products supplied by reputable businesses, preferably locally sourced and made. Well-rotted cow manure is one of the best because cows chew and grind their food more than other herbivores such as horses, so cow manure has fewer weeds and less urea, a source of nitrogen that, in large amounts, can damage your plants. Cow and horse manure are two of the easiest types of manure to get hold of in bulk, especially if you're able to take a drive out to the countryside, as farmers and horse owners tend to package the poop and pile it up outside their front gate. Grab a bag on your way home for a few bucks and your garden (and local community) will thank you later.

Unfortunately, I do have to add one caveat about manures: be careful where you get them. With so many herbicides being used these days, gardeners risk buying manure that contains glyphosate, which can inadvertently poison your plants. If the animals' feed – grass, grains or hay – is contaminated with a herbicide, the chemicals will flow through the digestive system and come out the other end! If this manure is then used on your garden, it can damage or even kill your plants. So if you notice dying plants with burnt leaves or 'whiteness' die-off (when the leaves curl and turn white or yellow) check to see if you've added manure to the bed because that could be the problem. Always get manure from a source you trust or test it first on a couple of plants before spreading it all through your vegetable garden or around your beloved fruit trees.

Digging weeds or a cover crop back into a garden bed and then letting it rest for several weeks before planting is another free and easy way to improve the structure and fertility of your soil. You can also make liquid fertilisers from weeds and other organic matter by soaking them in a water container to create 'compost teas'.

Using commercial fertilisers is okay as long as they're organic. Synthetic fertilisers are almost unavoidable in commercial agriculture, but never, ever use them in your home garden because they'll increase the salt in your soil and harm your plants. Synthetic fertilisers do nothing to add structure to your soil. In contrast, natural substances like good quality manures add extra fill to the earth and feed or grow worms, microbes and beneficial fungi. Trace elements or plant tonics like seaweed concentrate can also help keep your plants and trees healthy and productive. Citrus trees are a case in point. They require specific minerals not found everywhere, so a feed of trace elements and tonics will help tremendously.

When it comes to how much fertiliser to apply, always follow the directions on the packet. If you're unsure how much to use, start off with just a little and see how your plants go over a few weeks. If they're thriving, hold off adding more fertiliser, but if they look a bit pale and stunted, add some more. Ensure organically sourced or foraged fertiliser is well-rotted or mature before sprinkling it around the base of your plants or it will burn the roots. If you dress a bed with a good helping of green/fresh manure, you must rest the bed for several weeks before planting anything.

Hügelkultur

Hügelkultur is German for 'hill or mound culture'. It describes a medieval way of cultivating food on a mound of dirt. I stumbled on the method after having an idea to save soil by placing logs, sticks and other types of organic garden waste in the base of a raised garden bed. I searched the interwebs to see if my crazy idea had any merit and was delighted to discover that people had been growing food in the same way for centuries!

The Hügelkultur method is simple. Instead of filling your whole raised bed with premium soil, layer your heavy organic waste material such as logs, twigs and sticks in the base. Then add materials that will break down faster on top, like paper, cardboard, leaves and other garden waste. Finally, cover everything with your premium garden soil.

Over time, the materials under the garden soil slowly decay, creating a natural ecosystem and becoming food for worms, microbes and fungi. In turn, this adds nutrients and moisture to the bed for your fruits and veggies. How good is that? The organic matter in the base of the bed also holds water more effectively, acting like a sponge, so you don't have to water your crops as much. Your raised garden beds will continue to improve as you use them and add more compost and fertiliser.

Watering your vegetable garden and fruit trees

I really don't need to say how important water is when it comes to growing food! In a nutshell, a lack of regular watering will seriously affect the health and production of your crops. Here's a great takeaway: *regularly watering your vegetables a little* is much better than *occasionally watering a lot*. As far as your fruit trees go, once they're established, consistent watering is less important and can be focused more on fruiting times, periods of low rainfall or scorching weather. Most vegetables on the other hand have a shallow root system that makes them susceptible to heat stress, which can trigger them going into 'survival mode'. Even if they don't die, production and taste will be affected, and that's the last thing you want.

As I've said, I prefer to water by hand. If you're short of time, a drip irrigation system is easy to make. You can buy inexpensive poly watering kits from garden centres and hardware stores or you can buy individual pieces that you can join with some poly pipe or hose to match any garden design. Connect the system to a tap timer, and depending on the season, set it to water daily for half an hour in the early morning or late evening. This simple system should sustain your plants throughout the day in normal conditions.

If you have plenty of time, as I mentioned before, get out and hand-water in the early morning and late afternoon/early evening when there's less evaporation from the sun. The problem with watering systems on a timer is they water your plants even when you've had rain or it's been overcast. This can result in overwatering which is not only a waste of a precious resource, it can make your garden beds saturated and heavy. There's another great benefit to spot watering by hand:

as you wander around, you can keep an eye out for diseased leaves or pests and pick off bugs or caterpillars as you go, addressing problems early before they get out of control.

Hand-watering is underrated because most people see it as a bit of a waste of time. But I find it therapeutic, particularly in today's fast-paced world. I use my 'wandering watering time' to reflect, meditate and admire nature. I encourage you to consciously take the time to enjoy your garden and the work that has gone into creating it because satisfaction is a big part of why we put in the effort!

Flowers in the vegetable patch

Speaking of enjoying your garden, planting flowers in your veggie patch is a great way to add extra colour and attract beneficial insects like wasps, bees and ladybirds that will help pollinate your crops and attack nasty bugs. Natural chemicals and scents from marigolds can even repel underground pests like nematodes, while nasturtiums attract bees and double as a salad crop with edible flowers, leaves and buds. Leaving herbs and veggies to flower instead of pulling them out is an excellent way to add beauty and entice insects and larger animals like birds to visit your garden. Insect-eating birds are vital on our property as they hunt down and eat pests like caterpillars. It's one of the main reasons we don't need pesticides. And letting plants flower means they produce seeds that can be collected and sown next season.

Growing food in small spaces

We have three acres of land on our property just outside Brisbane, and while this is considered small by acreage and commercial farming standards, it's still much bigger than a standard urban block. Most people live in inner city and urban areas on small lots, or in apartments without a front or backyard, or in a rental home with restrictions on landscaping, so what can you do if you still want to grow vegetables, fruits or herbs?

Growing in containers, pots, vertical towers and wall gardens, grow bags and raised planters (with a base) have become popular in recent years. I've seen many outstanding examples of food gardens in small spaces, and even though I have plenty of room on our property, I still like growing some things in containers. The advantages are portability, minimal setup costs, ease of maintenance, and they can make a boring area look great! A courtyard or balcony garden with appropriate sunlight can become surprisingly productive, green and

inviting, and a fun challenge too.

Through my experiments, I've shown that you can 'crowd-grow' several different plants in one relatively small container. For example, using a 40-centimetre (16-inch) wide plastic pot, you can position a tomato plant in the middle with various herbs around the outside like basil, chives and oregano for an Italian mix. In another pot, you can plant a chilli in the middle with ginger, coriander and bok choy around the outside, and you have the base for all sorts of Asian dishes. This way, you can have these cuisines in several containers to grow and use for various meals during the week, and it barely takes up a few square feet of space.

Potting mix or the medium used in your container should be of good quality for best results. You need a different 'soil' than you would if growing inground or in a large, raised bed with an open base. Regular garden soil will quickly compact in a small pot affecting drainage and making the root zone anaerobic (lacking in oxygen), which plants hate. Just buy a cheap potting mix from the nursery and add some fertiliser or make it from scratch using homemade compost mixed with other ingredients to save money. I prefer to buy a premium potting mix with added slow-release fertiliser because it saves time, and I know I'm getting a consistent quality product for my plants.

I encourage you to start small. After that, you can think about Empire-building! There's nothing more demoralising than working hard and building a vast, expensive array of garden beds, only to realise that you neither have the time, the knowledge nor the ability to put in the effort to maintain it. A well-kept vegetable garden looks beautiful but will quickly become an eyesore if left to decay into a weedy paradise for snakes and spiders. If you have only one garden bed or several medium-sized containers, you'll be more likely to put in the effort because it doesn't take much time to look after. When you start to see the results and become more confident, it will motivate you to build more gardens or plant more containers as your skills as a food gardener grow.

How long does it take to set up a raised garden bed?

The good news is, you can have a high-sided raised garden bed ready (including planting and sowing) in less than half a day! An Aluzinc steel oval-shaped raised bed 2.5 by 1.5 metres (about 8 foot by 5 foot) takes about 30 minutes to screw together and then about an hour to fill. This is plenty of space for a beginner food gardener to start. You could grow around 30 large heads of lettuce or 10 tomato plants, or 18 cabbages, cauliflowers, broccoli or several dozen medium-sized onions.

A raised bed is also easier to net from pests and animals like possums. Just use a standard 25-millimetre (1-inch) irrigation pipe hooped over some bamboo stakes to make the framework and some standard bird netting secured over the frame with a few cheap clamps. Depending on how hard you work, you should complete your garden bed with netting in two to four hours and harvest your first salad in about three weeks.

When you start out, grow the food you would regularly buy at the supermarket. One of the aims, after all is to cut down on those grocery bills. As I mentioned earlier, there's no point growing beautiful beetroot, capsicums or Brussels sprouts if you don't usually eat them! (Unless of course you want to give them away, which can be an excellent idea too.) Brainstorm with other members of your household about your favourite veggies, salad greens or fruits and start with them first. Then you can branch out, growing other things you don't often eat or haven't tried before. There are more than 20,000 different types of tomatoes, so if you are only used to a few generic supermarket varieties, there are so many more to try! While you'll never be able to grow them all, you can do what I do: every year, I plant a few new varieties of my favourite fruits and veggies. Sometimes, I'll seek out varieties I've never heard of. Some of them have now become favourites I cultivate all the time.

It's exciting growing things you can't buy at the supermarket. Supermarkets priorities often aren't the taste but how good the product looks or how well it travels and will last on the shelf. There's not a lot of imagination to put it simply because the commercial process is such a beast that the supermarkets can't afford to be too experimental. Farmers supplying to chains are trying to make a living, so the more straightforward and cost-effective the process, the better. The downside of course is a lack of diversity. At home, though, we can afford to be more diverse and try new things. Many of us avoid unfamiliar veggies, especially children, so it's essential to shake up that pre-conditioned way of eating and try something different. When you break out of that mindset, a new and thrilling world of food is waiting for you to grow and eat.

So what are you waiting for? Get into it and get out gardening to grow yourself a tonne of organic home-grown food!

My top 100 food crops – listed in five categories of the best 20

I've chosen these five categories based on decades of life and gardening experience: 20 top vegetables (with a few 'fruitables'), 20 top fruits and fruitables, 20 top herbs and spices, 20 top unusual food crops and 20 crops I grew so you don't have to!

The aim of this top 20 format is to give you an easy-to-reference list of edibles with a simple Easy, Medium, Hard to grow value and some expert nutritional information so that you'll be enticed and inspired to give them a go yourself.

I understand that not everyone reading this book will live in an area where they can grow everything in this top 100 but I hope you can appreciate the principles and logic behind my choices.

And yes, I do know there's no such thing as a 'fruitable' but there is now because I just made it up! So what is a fruitable? Basically, it's a food crop that can be used as both a fruit *and* a vegetable. Typically, the seeds are grown on the inside of the edible part, for example, a tomato or cucumber, as opposed to a cabbage or carrots.

Lastly, these lists of top or best 20 are not in any order so don't overthink them too much as it's simply my opinion.

20 best vegetables

We live in a world where information about everything is literally at the tip of our fingers as we scroll through our social media and Google or AI our way to find out anything we want within seconds. I think this is a great advancement in technology but the downside is our susceptibility to exaggerated stories, unhealthy trends and outright lies.

One of the biggest and most misleading social media trends over the past few years I've noticed, is the lie that vegetables are bad for us. I understand that many of these posts are designed to get 'clicks' and shock people but there is a considerable movement of persuasive and influential people who are pushing the notion that humans are meant to only eat meat.

This belief is not only manifestly untrue it's also dangerous because it can lead gullible people down the 'garden path' towards a purely carnivorous diet that can be potentially life threatening.

Vegetables are extremely good for us and when eaten sensibly in a balanced way with animal proteins and other foods they play an important role in keeping our body and mind healthy. In essence, veggies fill in the nutritional gaps in our diet including providing the fibre, phytonutrients and antioxidants we need to fight off diseases like cancer.

The following 20 vegetables have been chosen because I love them – but also because of their nutritional value.

1. Kale

Easy – Very hardy and seeds germinate easily

Why?: Because I say, 'Hail the kale!' The reason I love kale so much is partly because of its consistent performance in the garden – and because it grows well without much care. Those big green-blue leaves look sensational. I've actually seen kale used as an ornamental plant in public gardens in Paris as it makes for a beautiful border plant.

Soil: Prefers well-drained soil however is quite forgiving – just add some compost.

Position: Full sun but will grow in part-shade.

Water: Regularly for crisp, plump and tasty leaves.

Top tip: A versatile plant that will grow in a wide temperature range. Will survive a frost, dry or even drought environment. Try fermenting kale with other vegetables and you'll be surprised at how good it tastes and what flavour it adds.

Uses: Finely chopped into salads, baked or fried for healthy, crunchy chips, tossed into a green smoothie, laid as a base for baked chicken dishes or other meats.

Nutritional power status: I hear all the time that if it tastes good it's probably not good for you, but in this case studies have linked consumption of kale to lower blood pressure and improved heart health. So munch and crunch your kale!

Brassica oleracea (Acephala Group)
**Curly kale, Tuscan kale,
Cavolo nero, Dinosaur kale**

Brassica oleracea (Capitata Group)
Green cabbage, Red cabbage, Savoy cabbage

2. Cabbage

Easy – Direct sow or raised in punnets

Why?: Cabbage goes back a long way in human history and was probably first developed in the Mediterranean thousands of years ago from crossbreeding kale varieties. Eaten raw, cooked or preserved through fermentation (sauerkraut) this vegetable is incredible!

Soil: Well-drained, rich soil with added compost.

Position: Prefers full sun where possible.

Water: Keep moist while germinating but be careful of overwatering as this can cause root rot.

Top tips: Feed with nitrogen liquid feed every 2–3 weeks for plump juicy heads and store in the fridge crisper for weeks if needed.

Uses: Fresh in Asian salads, coleslaw and stir-fries or ferment as sauerkraut. Use instead of a wrap, like a cabbage roll, pairing well with chicken or beef mince dishes.

Nutritional power status: No single vegetable can cure every disease, but there's a lot of scientific research into the benefits of eating cabbage. It eases inflammation, helps with nasty LDL cholesterol and is a prebiotic to feed healthy gut bacteria for a strong microbiome. (You might pop-off a bit more after some cabbage . . . but that's a good thing!)

3.   Lettuce

Easy – Shallow root system and fast growing

Why?: I'm not aware of many people who don't like lettuce for that added crunch in a sandwich or burger or in a big Italian salad bowl drizzled with olive oil and a splash of white wine vinegar.

Soil: Prefers free-draining soil enriched with manure to increase the nitrogen level.

Position: Full sun but will comfortably grow in part-shade especially in warmer climates.

Water: Keep moist until germination in about seven days and then water regularly as seedlings.

Top tips: Grow it fast with regular watering and soluble fertiliser every 3–4 weeks otherwise it becomes bitter and dry. Once it has gone to seed, harvest the seeds to grow next season.

Uses: In salads, as lettuce cups for mince dishes or San Choy Bow or pair with prawns for a 1980s prawn cocktail – a favourite in our house! Swapping out your fajita or burrito wrap for a lettuce cup adds a new twist to your meals, increasing the vegetable matter while lowering extra carbohydrates.

Nutritional power status: Essentially water and very low in calories, lettuce is great to fill out a plate to make you feel fuller for longer.

Lactuca sativa
**Iceberg, Cos (Romaine),
Butterhead, Loose-leaf**

Daucus carota subsp. sativus
Good varieties are Nantes, Purple Haze and Paris Market

4. Carrot

Medium – Due to the extra soil preparation required

Why?: If it's good for Bugs Bunny then it's good for us! Seriously, carrots are versatile veggies and a fun crop to grow. There's something really satisfying about pulling up a carrot buried in your garden.

Soil: Very loose, well-tilled and open type soil allowing for the large root to grow down unimpeded by compacted soil or stones etc.

Position: Full sun for best results.

Water: Keep moist, but not wet, until germination. Don't overwater as you will rot the root system.

Top tips: Sow them directly where you want them to grow as they do not transplant well.

Uses: Can be pickled, dried, roasted with honey and sesame seeds, blended into a soup or baked into scones, cakes or scrolls. You can also eat the carrot tops chopped finely as a tabouli-style salad.

Nutritional power status: Crammed with beta carotene for healthy eyesight. Also one of my top vegetables for insoluble fibre for a healthy gut. The carrot is very resilient in the lunchbox on the way to school or work and is great for the trip home if still uneaten.

5. Beetroot

Easy – Several seedlings can germinate from just one seed

Why?: Beetroot is another multi-use crop with the tops and bulbs/roots both edible. Some people hate it because they've only ever tried canned beets, but freshly harvested beets roasted in the oven like a potato are a totally different ballgame. Give them a go!

Soil: Well-drained and fertile for best results.

Position: Prefers full sun but will withstand part-shade.

Water: Water regularly every second day or as required depending on your location and soil moisture holding capacity.

Top tips: You can use the bulb and the leaves in a salad and the juice as a food colouring. Harvest young beets for a more tender texture. If you 'crowd grow' you can thin out for tender baby beets and leave the rest to grow full size.

Uses: This superfood can be roasted, blended and mashed or dried. Make purple pickled eggs or pink pasta for kids. Add to a health smoothie or as an ingredient in a popular poke style bowl.

Nutritional power status: Strong antioxidants, helps boost blood flow for exercise and has been shown to lower blood pressure. I love the colour pink too, which quite often can colour your poop! Nothing to worry about though.

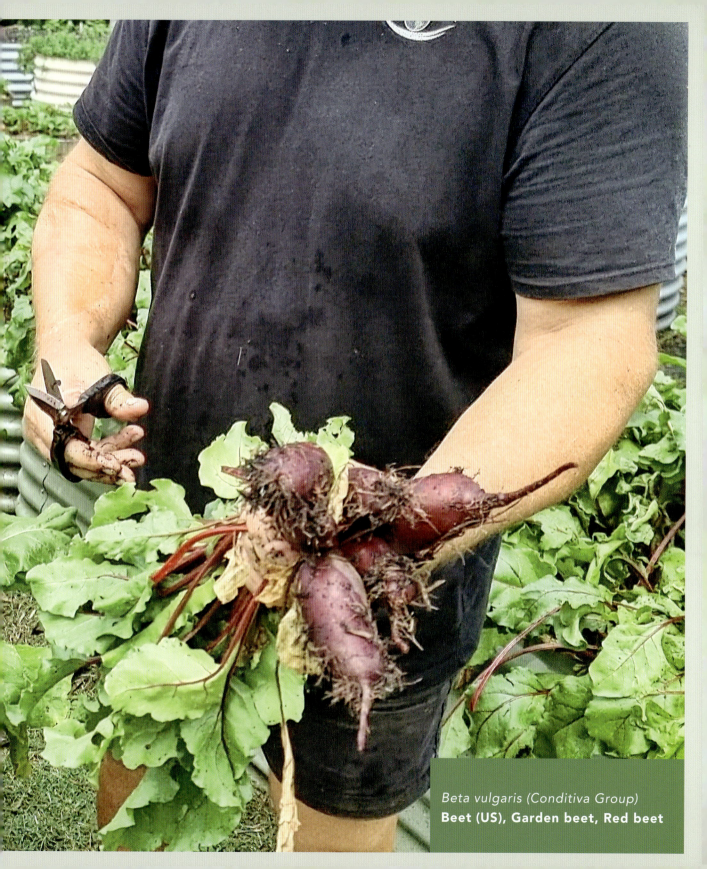

Beta vulgaris (Conditiva Group)
Beet (US), Garden beet, Red beet

Allium cepa

Brown onion, Red onion, White onion, Spring onion (green onion)

6. Onion

Medium – Sowing at the right time is key

Why?: Onions are tough to eat like an apple (yes, I've tried to more than once) but their ability to bring out the best in other foods is where onions shine. The base of a curry, melted into a slow-cooked sauce for pasta, caramelised on a hotdog or eaten raw with a salad or in a burger, their importance in food worldwide can't be overstated.

Soil: Free-draining soil due to susceptibility to bulb rot.

Position: Full sun.

Water: Regular watering, but don't overwater.

Top tips: Try planting onions that have a 'daylength' quality more suitable to your growing season and climate.

Some onion varieties grow better when daylength is long (up to 15 hours) or short (around 10 hours) or experiment with 'day-neutral' varieties. Also try spring onions because they are easier to grow than larger bulbed varieties.

Uses: Eat raw, pickled or caramelised. Eating them like an apple may not be to everyone's taste! Use to flavour foods instead of reaching for the saltshaker.

Nutritional power status: Great for making you cry, even greater for your gut and heart health. Their versatility in the kitchen makes them an easy contribution to your five serves of veg. Onions can add to your 'wind' production, so if you are on a long-haul flight or car trip, maybe reduce the onion load until you arrive!

7. Pea

Easy – As shelling peas . . .

Why?: From pea mash with fish and chips in England to pea and ham soup in ancient Greece every country has their own way to enjoy this amazing veggie whether it's a shelling variety or the type you eat pod and all.

Soil: Free-draining, compost-enriched soil.

Position: Full sun. A trellis is needed as they climb quickly and will fall over if not supported. If space is an issue, try a dwarf pea.

Water: Water regularly. Don't let dry out as peas can become bitter and stressed.

Top tips: Can be used as a green mulch for other crops when the growing season is finished. Peas fix their own nitrogen into the soil so leave roots in the ground after the crop is over for subsequent plants to utilise.

Uses: Dry roasted with wasabi as a snack (some commercial pea-snacks contain MSG and flavour enhancers, which I don't think are necessary and mask the real flavour of peas).

Nutritional power status: They may be small, but peas pack a punch. Vitamins C and E, carotenoids and zinc , to name just a few; all helpful for healthy digestion, eye health and immunity.

Pisum sativum
Garden pea, Snow pea, Sugar snap pea

Phaseolus vulgaris

Green bean, French bean, Bush bean, Climbing bean

8. Bean (legume)

Easy – and sometimes magic . . . Just ask Jack

Why?: Beans are almost a complete food on their own and contain most of the goodness needed for us to survive so they are great for self-sufficiency! They're also easy to store canned or dried to enjoy later.

Soil: Free-draining, compost-rich soil. No need for too much fertiliser. It grabs nitrogen from the air and stores it in the roots.

Position: Full sun.

Water: Regularly water and don't let dry out.

Top tips: Try different varieties all sown together. See which one flourishes in your part of the world or what type of bean you love the best.

Uses: Once the bean has matured let the seeds dry out. These can be regrown next season or stored dried in the pantry to be used in a stew.

Nutritional power status: Great source of protein and fibre which can help regulate blood sugar and reduce risk of type 2 diabetes. Being part of the legume family makes them an excellent replacement for meat, chicken or fish. They can be added to stir-fries, made into patties or fritters or blended down to a paste. They're also a good source of iron for blood flow and zinc for hormone building and healthy skin.

9. Potato

Medium – Some experience is helpful to grow well

Why?: Potatoes are one of the world's main food staples because they taste great and are full of calories. Populations that can grow potatoes won't go hungry. Plus, I don't know anyone who hates hot chips (fries) or mash with gravy for Sunday lunch roast.

Soil: Crumbly well-drained soil that's easy to dig.

Position: Full sun.

Water: Limited watering required until you can see sprouting or the plant is coming through the soil. Don't overwater otherwise the tuber will rot.

Top tips: Once harvested keep in a cool, dry and dark place such as a double-layered hessian bag (which mimics their life in the soil). Sunlight causes eyes to develop and a more fibrous potato, as well as a green compound called solanine which is poisonous.

Uses: Boiled, baked, thin sliced and baked, cubed for salads, added to breads and fish cakes for bulk. Mashed potato makes a fabulous mound for lamb shanks or savoury vegetable stacks to sit aloft.

Nutritional power status: More vitamin C than some fruits! That's one of the potato's little hidden secrets, however when you fry them into chips you do destroy some of the vitamin C. Cooked and cooled, they unleash more resistant starch and help support a healthy gut.

Solanum tuberosum
**White potato, Red potato,
Kipfler, Russet**

Ipomoea batatas

Kumara (NZ), Orange sweet potato, Purple sweet potato

10. Sweet Potato

Easy — Especially in warmer climates

Why?: Just like regular potatoes, sweet potato is a staple crop that has sustained many cultures in the past as a base vegetable. Unlike potatoes, sweet potato thrives in hot and humid conditions. It's also a tasty and versatile veggie with both the tubers and leaves able to be eaten.

Soil: Well-drained, fertile soil. Loose rather than heavy soil is best.

Position: Full sun to part-shade.

Water: Water regularly for a juicy tuber.

Top tips: Don't let the tubers get too large as they become woody and fibrous. The small to medium size have better flavour and taste.

Uses: As you would use potatoes, in casseroles, baked, as chips, to thicken stews or soups. Because it's sweeter than potato, sweet potato is fabulous when added to scones, muffins or cakes, replacing some of the flour in recipes.

Nutritional power status: Rich in beta carotene and potassium, they support eye health and promote healthy muscle and nerve function. The Okinawans in Japan ate a lot of sweet potato post-war when there was a food shortage. They're now part of their everyday diet and come in a number of colours. Sweet potato may contribute to their longevity too!

11. Corn

Easy – Even children can grow it (children of the corn)

Why?: One of the only grains that is worth growing at home because it produces a good amount of produce for the space. Commercially grown corn and popcorn are riddled with chemicals so growing our own helps limit our exposure to these nasties.

Soil: Free-draining, fertile soil. Add some compost. Don't overfeed with fertiliser as you will get 'leggy' plants with a long, thin stem that is not strong enough to hold the ears.

Position: Full sun.

Water: Regularly especially when flowering and fruiting.

Top tips: Bury the husks and unused cobs in the garden for worms and microbes as they love them. Grow in a protected spot, away from high winds as they are likely to blow over. Use a trellis or stakes with a stringline to stabilise the plants.

Uses: Grind the dried kernels to make corn meal/grits. A handy gluten-free alternative to wheat flour.

Nutritional power status: Gluten-free means corn is fantastic for those with coeliac disease or gluten intolerance, and high in lutein and zeaxanthin which protect against retinal damage. Boiling or microwaving brings out the lovely, sweet flavour, or try chargrilling on the BBQ for a more 'smoky' flavour.

Zea mays
Sweet corn, Field corn, Maize

There are several types of corn but the main ones you should know as a backyard grower are:

Sweet Corn (*Zea mays saccharata* or *saccharinus*)
Which is the most known variety grown to cook and eat normally (like corn on the cob).

Popcorn (*Zea mays everta*)
Which has a hard kernel with trapped air inside that explodes when heated/cooked in butter or oil making that wonderful, popped corn we love to eat at the movies.

Flint Corn (*Zea mays indurata*)
Hard corn that isn't very nice to eat the standard way, however, we can grind it down to use as polenta, grits or a wholemeal kind of cornflour to make tortas etc.

Dent Corn (*Zea mays indentata*)
Has a "dent" in the top of the kernel when dried and is mainly used for animal feed and processed oils for food. Growing dent corn is probably a waste of time for most people as you can feed the other types of corn to animals anyway, but the other three are worth giving a go!

Coloured dent corn

Popping corn

Sweet corn

Brassica rapa (various groups)
Bok choy, Choy sum, Tatsoi, Mizuna, Chinese cabbage

12. Asian greens

Easy – Grows just about anywhere at any time

Why?: The clue is in the name. While rice plays an important part in the Asian diet, Asian greens like bok choy, mizuna, wombok, tatsoi and mustard greens are arguably just as significant because it's these vegetables that help flavour and add nutrition to the oriental meal. In the West, we often eat rice with meats but in the East meats are expensive and scarce (even today in many places) so greens are a healthy substitute.

Soil: Free-draining soil.

Position: Full sun to part-shade.

Water: Water regularly and feed with a liquid fertiliser containing nitrogen every 3–4 weeks.

Top tips: As these greens are very fast-growing harvest the leaves or plants regularly to prevent going to seed. Succession sow or plant out seedlings every four weeks over the growing season to extend the harvest.

Uses: Stir-fry, salads and soups or use bigger leaves to wrap your favourite protein or vegetable parcel.

Nutritional power status: Rich in folate, iron, magnesium and multiple vitamins, they help to prevent chronic disease development. Medical specialists recommend increasing folate during pregnancy to reduce the risk of birth defects.

13. Asparagus

Medium — Needs care and is a perennial

Why?: Once established, asparagus will give you tasty spears from the one plant for up to 25 years! Yes, there is some seasonal maintenance but it really is one of the few vegetables you can plant and 'forget' and then continue to harvest for decades.

Soil: Pretty forgiving, but make sure the growing medium is not too heavy. Asparagus has a long root system so allow at least 50 centimetres (20 inches) of soil depth for best results.

Position: Full sun to part-shade.

Water: Water regularly early in the season to encourage fast-growing juicy stems (spears) or, if establishing your plant, especially from seeds, until the crown and root system is developed.

Top tips: Mulch back on itself by cutting down the old dead ferns and adding extra mulch in winter. Top up the bed with animal manure or organic fertiliser in spring before it starts sprouting. It will take 2–3 years before first harvest and 2–3 cycles for the spears to grow out and die off before you get a good harvest of thick spears.

Uses: Their unique flavour and texture make them perfect additions to salads, sushi, risottos, grilled or pickled.

Nutritional power status: Good for heart and bone health due to high potassium levels, and high fibre content fuels the good bacteria in your gut. You may notice your pee has a 'funky' smell to it after eating asparagus. This is due to an acid called asparagusic, which is broken down into sulphur containing by-products. Supposedly not everyone has the ability to smell this though.

Asparagus officinalis
Sparrow grass

Brassica oleracea (Italica Group)
Colewort or Field cabbage

14. Broccoli and broccolini

Easy – But can take up a lot of space

Why?: I like the taste of it! Also the plants are highly ornamental with those big leaves and tight green heads. What a wonderful example of how cross-breeding and selection over thousands of years can help create such a top veg.

Soil: Free-draining, fertile soil. Add nitrogen prior to planting and give regular liquid feeds for plump heads.

Position: Full sun.

Water: When necessary. Don't let the plant dry out as it will stress and possibly open the flower prematurely going to seed before it grows big enough.

Top tips: Harvest when the heads are tight and before the florets open as once the tiny buds begin to flower it will bolt to seed very quickly!

Uses: A great, green leafy addition to any roast meal. Fried with soy sauce and bacon makes it a stand-alone side dish. Equally tasty is a cold broccoli salad. Cook till just al dente and then plunge into an iced water bath to blanch. Broccolini is a cross-breed of broccoli and Chinese broccoli, a leafy vegetable commonly referred to as gai lan in Cantonese or jie lan in Mandarin. More mellow and less bitter than broccoli and the texture of the stem is softer.

Nutritional power status: From the brassica family, an excellent fibre source! Surprisingly high in vitamin C to fight off cell damage.

15. Cauliflower

Easy – But never a substitute for real steak

Why?: Because cauliflower steaks are to die for! I'm joking . . . my favourite way to eat cauliflower is baked in a cream sauce topped with melted cheese – crikey, that's yummy!

Soil: Free-draining, fertile soil. Add nitrogen. Requires regular liquid feeds for plump heads.

Position: Full sun.

Water: When necessary. Don't let the plant dry out as it will stress (like broccoli).

Top tips: Consider growing purple cauliflower for a pop of colour in your meals and the wow factor on the table or in fermented preserves.

Uses: Instead of rice for a low carb option, mixes well into curries, flash-fried and mixed with a garlic tahini dressing is delicious. Cauliflower patties are a great alternative to a meat patty on a burger.

Nutritional power status: Contains nearly every vitamin and mineral that we need, and research shows promising evidence towards fighting certain cancers.

Brassica oleracea (Botrytis Group)
Green cauliflower or Broccoflower

Cucurbita maxima / pepo / moschata

Butternut, Kent, Queensland Blue, Jap, Sugar pumpkin

16. Pumpkin

Easy – However, it does require a lot of space

Why?: Pumpkins are one of those 'fruitables' I mentioned earlier but I've included them in the vegetable list instead of the fruit list because it's a cucurbit (or gourd) that's usually cooked before eating (unlike cucumbers). Pumpkins can be stored for months after harvest and one plant can potentially grow many large fruits making it an extremely productive crop.

Soil: Very forgiving, strong rooting system. Grows best in fertile soil with lots of compost.

Position: Full sun to part-shade.

Water: You can tell when it needs water by looking at the leaves. When they wilt, up the water. Watch out for powdery mildew though.

Top tips: It's a hungry plant but if the vine is left to wander and root down in other places as it goes, it will produce a lot of large quality pumpkins without the need for much fertiliser (if the soil below has decent fertility). If grown in a restricted space, give plenty of fertiliser and limit fruit to a few for best results.

Uses: Roast the seeds, baked with the skin on and sprinkle with salt. And everyone knows what an excellent soup it makes.

Nutritional power status: Skin on or off, pumpkin is a great source of fibre. Bonus of beta carotene (a pro-vitamin A molecule) makes it an awesome vegetable to mash, roast or boil.

17. Zucchini/squash

Medium — Can be susceptible to diseases and pests

Why?: Also a fruitable, zucchini and button squash are a delightful side dish when simply fried in butter and seasoned with salt and pepper. They are essentially a smaller, faster-growing and thinner-skinned kind of pumpkin. As I always say, 'Put on your boots and grow some zuccs!'

Soil: Free-draining, fertile soil.

Position: Full sun.

Water: Water as needed during the growing season and try to water the base of the plant rather than the leaves to limit fungal diseases spread by damp foliage.

Top tips: Harvest small to medium-sized for a tender fruit and grow some to large size to harvest the seeds for the next season. They can also be grown up a stake or trellis for easier harvesting, keeping the fruit away from snails and slugs.

Uses: Great in stews and soups as minces, and blends well adding bulk to meals. Use large fruit as a mince pot or boat. Scoop out a zucchini longways and add to leftover bolognese. Scoop back into zucchini, top with cheese and cook till soft and cheese has melted. Yummy!

Nutritional power status: Low calorie but high in fibre, zucchini is versatile. It contains many B vitamins that help to improve brain function.

Cucurbita pepo

Courgette (UK), Summer squash, Patty pan squash

Raphanus sativus
**Daikon, Cherry belle,
Watermelon radish**

18. Radish

Easy – And the fastest-growing veggie

Why?: Some people don't like the pungent flavour of radish but instead of disregarding this veggie they could try a mild variety such as daikon (Japanese radish). I love eating all varieties fresh, pickled or fermented.

Soil: Well-drained, fertile soil.

Position: Likes full sun but will do okay in a pot in the shade.

Water: Regularly to keep the bulb crunchy and prevent woodiness.

Top tips: Grows very fast so try succession growing rather than one bulk planting. Sow a little at a time one week apart to stagger the crop.

Uses: Eat small to medium (except for daikon), Chinese or Asian pickled radish in vinegar and sugar, roasted, stir-fried, in a salad or as a condiment with cheese. Experiment with all the varieties of colours as some are spicier than others.

Nutritional power status: Packed with powerful antioxidant properties, radish has been shown to help reduce the risk of colon cancer.

19. Spinach

Medium – It can be finicky and doesn't like hot weather

Why?: I thought it would give me big muscles. But now I eat spinach for the mild flavour and only grow it during the cooler months in our subtropical climate.

Soil: Free-draining, fertile soil.

Position: Full sun to part-shade.

Water: Regularly, don't let dry out or wilt, it will become bitter and soft instead of plump.

Top tips: Harvest leaves regularly to encourage new growth.

Uses: Quiche, spinach pie, omelette, rolled with soft cheese and chicken or pork rolls.

Nutritional power status: Pair with egg as the vitamin C unlocks the iron from the egg, and an easy vegetable to sneak into pastas, smoothies and stir-fries. Also great raw ingredient to add to breakfast smoothies or protein shakes.

Spinacia oleracea
English spinach, Baby spinach

Apium graveolens
Stalk celery, Pascal celery, Celeriac

20. Celery

Medium – Some management required to grow well

Why?: Celery sticks and dip, soups, and my wife Nina loves it juiced with other ingredients to make a healthy drink.

Soil: Free-draining, nutrient-rich soil with added compost and soil minerals.

Position: Prefers full sun.

Water: Regular watering especially during germination and establishing after planting out seedlings.

Top tips: Blanch the stems by wrapping in cardboard or using PVC piping as this will grow a white, tender and sweeter stem. Regularly check for slugs and snails and pick them off.

Uses: A great green leafy addition to any roast meal, soup or casserole. Good to add to a platter for dips or munch on a stick if you need a quick snack!

Nutritional power status: A nutrient-rich, low-calorie, high-fibre food. Two stalks of celery contain only 15 calories. Aside from its low calorie count, it contains dietary fibre and vitamin K for blood clotting. Also 95 per cent water so a very hydrating vegetable.

20 best fruits and 'fruitables'

The first four in this top 20 list are fruitables because they are technically a fruit but are often used as or mistaken for a vegetable.

One of the main reasons why I grow our own fruit (and fruitables) is so that I know what chemicals are in and on the produce – in our case, absolutely none! Unfortunately, the same can't be said for what we buy from the supermarket or what's served to us at a restaurant or fast-food outlet. It's taken a long time for regular consumers to catch on but at the time of writing (2025) I think people are finally realising the faith they have placed in authorities, big agriculture, pharmaceutical companies and the food industry to keep them safe from harmful chemicals is unfounded.

I genuinely hope the new trend of openness and scepticism continues to grow and forces positive changes in how our food is farmed and processed in the future. In the meantime, we can limit our exposure to harmful chemicals by growing as much fruit at home as we can. With the number of dwarf fruit trees now available in almost all species you don't need a big property to grow your favourite fruits.

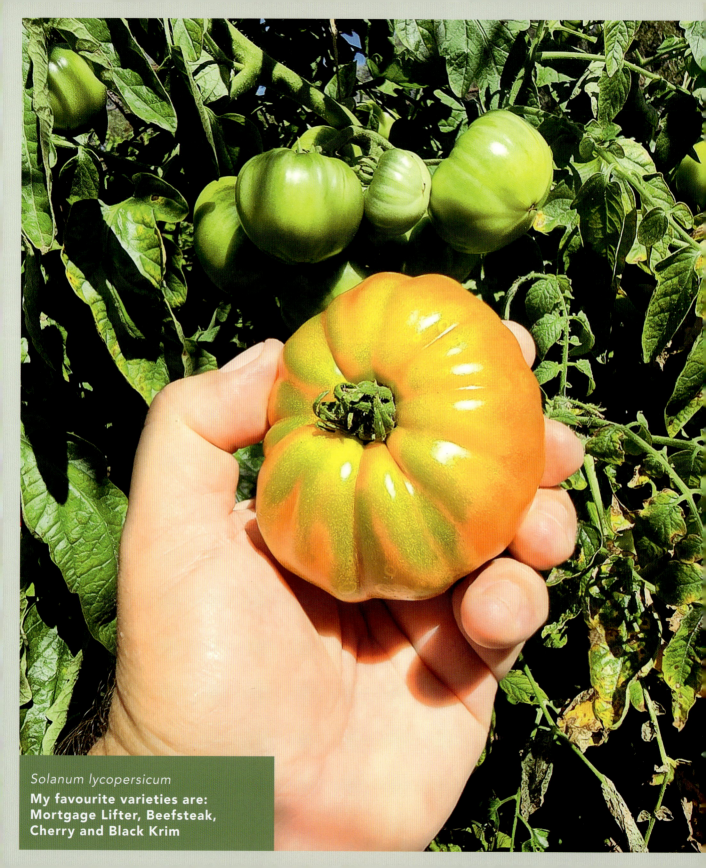

Solanum lycopersicum

**My favourite varieties are:
Mortgage Lifter, Beefsteak,
Cherry and Black Krim**

1. Tomato

Medium — Larger varieties can be challenging due to pests and disease

Why?: Tomatoes are a fruitable and my favourite food. I'm not just saying that. If I were to pick one food to rule them all it would be toms because I love the flavour, and they can be used in so many ways. We even travelled all the way to Spain to partake in the La Tomatina festival and what a savoury experience that was!

Soil: Very fertile and free-draining. Mixing in some animal manure and organic fertiliser before planting will get your tomatoes off to a great start.

Position: Full sun where possible but will grow in part-shade particularly, the cherry varieties.

Water: Important to water regularly, especially when the fruit is nearing maturity as intermittent watering or letting the plants get too dry between drinks can cause the fruit to expand and the skin to crack.

Top tips: Water at the base of the plant to limit fungal disease – keep the leaves dry. Prune foliage to allow air to circulate around the plant. Blossom end rot means the soil lacks calcium; adding some crushed eggshells or garden lime to the soil will help to return the calcium. Start with cherry tomatoes and then progress to the beef steak varieties which are a little harder to grow.

Uses: Sauces, pastes, chutneys. Tomato toasties (tomato and cheese melted on toast) is a breakfast favourite of mine since I was a kid.

Nutritional power status: Crammed with lycopene, which has shown that it may help protect against cancer such as prostate and cardiovascular disease. An easy addition into salads, wraps or sandwiches to get that vitamin C. One of the only foods that increases in nutrition after cooking.

2. Cucumber

Easy – As cool as a cucumber

Why?: Cucumbers are a fruitable and my favourite pickle is fermented dill pickled cucumbers. I'm a big believer in eating fermented foods for good health. I might not understand why completely (no one does) but my gut feeling is that good gut health is incredibly important for overall body health.

Soil: Free-draining soil with a good mixture of rich organic matter.

Position: Full sun.

Water: Regularly to form crispy fruits.

Top tips: Know your variety and make sure you pick cucumbers at the appropriate size. Most cucumbers are best harvested at around 15 centimetres (6 inches) long otherwise the flesh may become bitter and hard or hinder the growth of other cucumbers. Use a trellis if the fruit is starting to rest on the soil as they can rot where the skin touches the ground.

Uses: In salads, ribboned, pickled (of course), juiced, and as a side to help cool down a spicy curry. You can use them placed in the eyes to rehydrate the skin. Sadly my wrinkles are too deep for that to work these days.

Nutritional power status: Eat your water! Include cucumber to keep you hydrated and prevent constipation. Have you heard of pickle juice? It's basically the fermented brine that the dill cucumbers were pickled in. This juice is known to quickly help athletes rehydrate and replenish their electrolyte levels to stop or prevent dehydration and muscle cramping.

Cucumis sativus
Gherkin, Garden cucumber, Khira

Capsicum annuum
Bell pepper, Paprika (Europe)

3. Capsicum/sweet pepper

Hard – susceptible to pests and disease

Why?: Another fruitable, capsicums might be difficult to grow but they are totally worth it because of how good they are to eat – and how expensive they are to buy.

Soil: Free-draining, fertile soil with added compost.

Position: Full sun.

Water: Regular watering.

Top tips: Cover plant or use insect mesh if insects or animal pests are prevalent in your area. Also ensure you sow and plant out at the right time for your climate to get good fruiting results as the plants are very sensitive to temperature.

Uses: Chargrill to produce a blackened skin. Place in a plastic bag while warm to sweat skin to make it easier to peel. Place it in olive oil or vinegar and use as antipasto. Chop and use in stir-fries or raw in salads.

Nutritional power status: Contains more vitamin C than an orange! Helps your body to fight off cell damage.

4. Eggplant

Easy – Especially, modern heirloom varieties

Why?: Melanzane sott'olio (eggplant cured in olive oil) is a traditional Italian preserved eggplant so good that you'll fight Nonna for the last piece on the antipasto platter. I can say proudly that I have mastered this recipe and everyone I know loves it. Eggplant is one of my favourite crops to grow because in our subtropical climate it has an extra-long growing season which translates into plenty of fruit. Oh, and eggplant is my last fruitable.

Soil: Free-draining, rich and fertile soil.

Position: Full sun.

Water: Regular watering because like tomatoes irregular watering can make the skin crack.

Top tips: Pick fruit at the appropriate size and don't let grow too big, otherwise the skin will crack and the flesh becomes bitter and woody.

Uses: In salads, stir-fries, casseroles or curries as it has a strong flesh that holds its form. Small to medium eggplants don't need salting. Salting larger eggplants improves the sponginess and produces a creamy texture. Try crumbed eggplant steaks or fingers for dipping into sauces or salsa. Use instead of pasta in Moussaka or layered Italian lasagne-style dishes.

Nutritional power status:
Great source of fibre and low in carbohydrates. The beautiful purple colour is rich in antioxidants, which are molecules that help the body eliminate unstable molecules that can damage cells if they accumulate in large amounts. Foods that contain antioxidants may help prevent a range of diseases. Among the antioxidants in eggplants are anthocyanins, including nasunin, lutein and zeaxanthin.

Solanum melongena
Aubergine (UK), Brinjal (India)

Persea americana
Alligator pear or Avocado pea

5. Avocado

Hard – For most backyard growers unless soil and climate are perfect

Why?: Smashed avocado on toast that's why . . . I reject the notion that only rich people and stuck-up university students should partake in this delicacy! The average price in Australia for a fresh quality avo is $3 and rising so a productive tree or two in your backyard is literally big green money balls swinging from branches.

Soil: Must be free-draining, deep, fertile soil (think rainforest floor type rich humus). A week or so before planting, add some blood and bone or manure mixed in with plenty of compost to improve the soil and get the young tree off to a good start. Avocado trees will *not* grow in heavy or clay soils so don't even try. Once the tap root of an avocado tree reaches clay soil it will quickly die. If you do have heavy soil, you will need to create a hill or mound of good free-draining soil at least a metre or 3 feet above the existing clay and plant the tree into that. Hopefully, this "hilling up" will prevent the tap root from growing into the clay and your tree will survive. Avocados can grow well in pots, just use a good-quality, free-draining potting mix and try a dwarf variety.

Position: Full sun or can grow in part-shade (in hot climates).

Water: There is a misconception that avocado trees don't like too much water, but it depends on the soil's water-holding qualities. If the soil is free-draining and oxygenated the tree will love the extra water, but if the soil is heavier, it's best not to overwater or phytophthora root rot may occur, and your tree will be toast (without avocado on it).

Top tips: The first three or four years of an avocado tree's life are critical. If you can get your avocado tree past this time, there's a good chance it will grow well for a lifetime. In hot weather the stem of a young avocado tree can suffer sunburn so placing some shade cloth loosely around or over the tree through the hottest part of the year can help.

Uses: Apart from adding to salads, use in omelettes and egg dishes, smoothies, sushi, guacamole, salsas and as a healthy fat ingredient in sweet cakes, slices and mousses. Cut it in half, remove the seed and fill the hole with white vinegar and a generous sprinkle of salt and pepper then eat out the avocado flesh with a teaspoon.

Nutritional power status: Healthy monounsaturated fats and fibre. They also provide many other vitamins and minerals and antioxidants. Much research has been conducted into the health benefits of avocados and reducing the risk of heart disease.

Citrus limon

Top varieties: Meyer, Eureka, Lisbon, and Bush lemon

6. Lemon

Easy – Lemon squeezy!

Why?: It's almost a sin not to grow one of the most famous fruits in the world! They grow easily in most climates and the uses for lemons are far broader than just a squeeze of juice over your oysters.

Soil: Lemon trees are quite forgiving and can grow in a range of soils.

Position: Full sun.

Water: Regular watering will ensure strong growth and juicy fruit.

Top tips: Fertilise at the beginning of each fruiting season. Use a citrus fertiliser with trace elements as they need specific minerals to grow healthily and produce top quality fruit. Try different varieties if you have the room such as Meyer, Bush Lemon and Eureka.

Uses: Use as a versatile flavour instead of seasoning. High in pectin so good for setting jams. Use peel or zest in stir-fries or baking. If you have an abundance of lemons in season, juice and pour into ice cube trays and store cubes in the freezer for months of lemon juice. Dried or dehydrated can be used to make refreshing sparkling water or a soothing tea. Preserved fermented lemons in salt (Moroccan-style) are added to dishes like curries and tagines.

Nutritional power status: When life gives you lemons, eat them! Full of vitamin C, squeeze some lemon over dark green leafy vegetables to maximise your iron absorption.

7. Lime

Easy – You'll never pay for a lime again

Why?: A few years ago we were visiting some friends in the city and the host was cutting up limes for the drinks. I noticed that one of the limes had a price sticker on it for $2.50. I said, 'You didn't pay that much for each lime, did you?' The reply was a prompt nod of the head and a 'yep'. At the time, we had so many limes on our one tree it was impossible to use them all so from that day onwards whenever we visit friends, we bring a bag of limes.

Soil: Like lemons – will grow in a range of soils.

Position: Full sun.

Water: Likes plenty of water after flowering and during the development of fruit.

Top tips: Lime trees can grow a little bigger than most lemon trees, even though the fruit is smaller, so make sure you give them plenty of space. Keep an eye out for signs of citrus borer which are sudden branch die-offs and deal with it quickly by finding the bore holes and cutting off the branch and burning or squashing the grub.

Uses: Experiment with different limes such as Tahitian for the juice or Makrut for their flavourful leaves or zest from the inedible fruit.

Nutritional power status: Flavoursome and tangy, lime is a great addition to your salad dressings to keep your immune system healthy thanks to their high vitamin C content.

Citrus aurantiifolia / Citrus latifolia
Key lime, Persian lime

Citrus × *sinensis*
Sweet orange

8.   Orange

Easy – And you can always depend on a Valencia!

Why?: To get the citrus trifecta, of course! If you're in a climate zone suitable for oranges this should be the first fruit tree you plant. Freshly squeezed OJ is still considered an indulgent treat so good luck getting a glass at your local café for under $10 these days.

Soil: Like most citrus, orange trees will grow in a range of soils.

Position: Full sun.

Water: Keep watering regularly, especially if you are in a dry part of the world. Oranges need plenty of water when fruiting as an orange is around 80 per cent water and you always want a juicy fat fruit.

Top tips: If you have the space grow different varieties that mature at different times to extend your harvest. For example, a Washington orange will ripen early in the season, Lane's Late Navel orange will ripen later in the season and a Valencia orange will hold on to the tree for weeks (sometimes months) after ripening for you to harvest them last. Our orange season typically lasts around 6 months of the year.

Uses: Oranges can be used in a range of meals from breakfast juice or orange peel in granola, to stir-fried orange segments with chicken or duck, to crystallised or candied orange in salads or the zest in a range of sauces. Use in baking to decrease the amount of added sugar.

Nutritional power status: An orange a day keeps scurvy away! Having one orange daily will meet your vitamin C requirements and will reduce inflammation in the body.

9. Plum

Easy — Even the dwarf varieties produce a tonne of fruit

Why?: Plums are delicious but like so many commercially grown fruits these days they are often picked prematurely for better transportation at the sacrifice of flavour. When you grow plums at home you can ripen them fully on the tree, so the sugars develop properly to give the fruit the best taste possible.

Soil: Plum trees will grow in a range of soils including clay.

Position: Full sun.

Water: Normal regular watering, but they need more during the fruiting months. They can withstand long dry periods making them a low-maintenance fruit tree.

Top tips: The fruit can be susceptible to pests such as animals, insects, birds and fruit fly. So after flowering, cover the tree with cloth, mesh, or individually bag branches of fruit for protection.

Uses: Apart from jams, jellies and conserves, make a sauce for savoury meat or vegetarian dishes. Experiment with green plums or try them salted.

Nutritional power status: Containing a bunch of antioxidants and fibre, plums keep you regular and make your gut microbiome and immune system happy.

Prunus domestica
Common plum, European plum, Plum, or Prune plum

Fragaria × *ananassa*
Good varieties: Albion, Florida Fortuna, Sundrench

10. Strawberry

Medium – Every garden should have a strawberry patch

Why?: Strawberries are one of the 'dirtiest' foods you can buy at the supermarket. So if you want pesticide-free strawberries you really must grow them yourself. There are several ways you can grow them such as in the ground with soil mounded up, in tubs, pots, vertical containers, hydroponically, and of course in raised beds.

Soil: Free-draining, fertile, slightly acidic soil. You can use chicken manure to increase the acidity of the soil but they should grow well in most soil or potting mixes.

Position: Full sun.

Water: Due to the shallow root system strawberry plants are susceptible to drying out especially in warmer climates so make sure they are watered regularly in hot weather. If growing in containers try a slightly heavier potting mix or add some extra compost to improve the water-holding capacity or you'll be forever watering them to keep the plants from stressing out.

Top tips: Add extra potassium fertiliser to increase the fruiting and sweetness of the fruit. Plant different varieties to extend the season and enjoy slight variations in flavour and size. Regrow next season's plants from runners sent out from your current crop by repotting them into small containers to mature in a protected area like a greenhouse.

Uses: Dry or dehydrate for breakfast cereals – use an oven on very low to slowly dry them out. Dipped in chocolate, they make a lovely gift or slightly healthier treat.

Nutritional power status: Sweet but mighty, strawberries are high in fibre, potassium and polyphenols which help to prevent heart disease. They are a super sweet treat fresh, frozen or blended with only 32 calories per ½ cup or the carbohydrate of only a third of a banana; great for diabetes and your sweet tooth!

11. Mango

Easy – In the right growing climate (tropical to subtropical)

Why?: Obviously, the fruit is amazing but the price you pay at the supermarket would lead most people to think mangoes are hard to grow when it's not the case. Mangoes grow so fast and easily in warmer climates that the hardest thing is keeping the tree from getting too big and out of control.

Soil: Mango trees are very hardy and happily grow in a range of soils, but free-draining, fertile soil is best.

Position: Full sun and good airflow to reduce fruit and flower loss through fungal diseases like anthracnose.

Water: Water well during the growing season to encourage juicy fruit.

Top tips: Mangoes are grown in mostly tropical climates but can still grow well in the subtropics with care and protection such as a hot house it's possible to grow them in cooler regions. The fruit is highly fragrant and attractive to animals so protecting your mangoes with bags or tree netting is usually required.

Uses: Eat straight from the tree or use in savoury salads or fruit salads. Use mangoes green in a leafy salad with sesame oil, chilli and lime juice for a tangy Asian taste sensation. If you have a glut of mangoes (lucky you) and can't possibly eat them all, mango strips can be dried, pureed or frozen for later.

Nutritional power status: The juicy combination of fibre, vitamins and potassium has shown to decrease the risk of heart disease by controlling blood pressure and cholesterol levels.

Mangifera indica
Varieties: Bowen (Kensington Pride), Glenn, Nam Doc Mai, R2E2

Pyrus pyrifolia
Asian pear, Apple pear

12. Nashi pear

Easy — and fast growing

Why?: Nashi pears don't go soft like regular pears and stay hard like an apple. But unlike a normal soft pear, nashis taste great when hard which makes them easier to manage for the home gardener.

Soil: Very forgiving however best in well-drained, rich soil. Can grow in heavy soil though.

Position: Full sun where possible, okay in part-shade.

Water: Regular water at the base of the tree.

Top tips: The developing fruit is a target for fruit fly sting such as Mediterranean or native Queensland fruit fly. If you live in an area that has fruit fly after flowering protect the emerging fruit with insect mesh over the whole tree or bags over the fruit.

Uses: Fresh, dried or pear puree can be used as a sauce, in baking or in a savoury pork stew. Layer on pastry for a simple dessert with a sprinkle of cinnamon.

Nutritional power status: Pears have a hidden secret weapon to move your bowels! They contain dietary fibre, including prebiotics, which promotes bowel regularity, constipation relief and overall digestive health. To get the most fibre from your pear, eat it with the skin on. High in sorbitol and fructose known as FODMAPs, which may produce excess gas for those with irritable bowel syndrome (IBS). If you don't mind a bit of flatulence then gnash a nashi!

13. Peach

Easy – And fast to establish

Why?: It's not easy finding a tasty supermarket peach these days because they are often harvested too early (for better transportation) before the sugars have a chance to develop. Grow your own and let them fully ripen to find out what a peach *should* taste like.

Soil: Another fruit tree that will grow in a range of soils from sandy to clay.

Position: Full sun.

Water: Surprisingly drought tolerant but regular water will produce the best fruit.

Top tips: Try growing different varieties: white, yellow or angel peach (the flat ones; an easy-growing variety, great for lunchboxes). To keep the tree easy to net consider espalier (training the branches on a structure to reshape the nature spread). Vulnerable to fruit fly strike so netting or bagging fruit might be needed.

Uses: Preserve, use in tarts, stews or simply eat straight from the tree.

Nutritional power status: Peachy keen to keep a sharp mind? Research has shown that the compounds in peaches can help prevent the risk of Alzheimer's disease.

Prunus persica
Varieties: Tropic Beauty, Angel, Flordaprince

Ananas comosus
Ananas, Pina, and Abacaxi

14. Pineapple

Easy – Just cut the top off

Why?: My father-in-law first showed me how easy pineapples were to grow by placing the top of one in the garden outside our back door. He didn't even bury it and to my amazement the spikey top took root and grew like crazy. Two years later we had another pineapple! Now we have dozens planted all around our property.

Soil: Grows in heavy to sandy soils and very drought tolerant.

Position: Full sun to part-shade and even dappled shade, although they will take longer to fruit than if in full sun.

Water: Prefers regular water however is hardy and resilient.

Top tips: Grows well in containers and makes an excellent low hedge garden border. Always pick when ripe which is when the stalk falls over and is not supporting the pineapple anymore. The fruit will not continue to ripen after picking so make sure it is ripe before removing it from the plant. Look for a sweet smell, bright colour with some orange (note that not all pineapples colour orange when ripe as some stay green) and the fruit should give slightly when pressed.

Uses: Fresh pineapple, healthy fruit and vegetable juices, fritters or with savoury meats such as duck, pork or chicken. Or simply throw on a burger or pizza.

Nutritional power status: Sweet and tangy, pineapple helps to fight off bacterial infections while also helping healthy bone formation due to the high manganese content.

15. Mandarin

Easy – The easy-peel varieties are the best

Why?: What do you call a mandarin that is hard to peel by hand? An orange. Look, I do love the flavour of our Honey Murcott mandarin, but I hate trying to peel them and for me a true mandarin is one that my big paws can skin in a matter of seconds.

Soil: Can grow in a wide range of soils.

Position: Full sun.

Water: As needed if natural rain is limited. Water more during fruit development or you might get early fruit drop or smaller than usual fruit.

Top tips: Grow an easy-peel variety such as Clementines that are small, bright in colour and have a thin easy-peel 'zipper' skin. We have an Imperial mandarin variety, which are easy peel and delicious to eat especially straight off the tree.

Uses: Eat fresh, not great as a juice as can be quite bitter (Honey Murcott juice is sweet and I'd rather juice them over trying to peel them any day).

Nutritional power status: Packed with vitamin C and soluble fibre, mandarins help to reduce LDL cholesterol and keep you feeling fuller for longer.

Citrus reticulata
Tangerine, Satsuma, Clementine

Cucumis melo / Citrullus lanatus

Includes Rockmelon, Cantaloupe, Honeydew, Watermelon

16. Melons

Medium – Can be challenging to grow

Why?: I love big juicy melons such as watermelon, rockmelon and honeydew melon and the many closely related varieties. There are other 'melons' that are more obscure and some of them get a mention later in the book. But when I think of melons I recall my childhood around Christmas time at my grandparents' home running around with my cousins in the backyard with a slice of juicy red watermelon in our hands. Those were the days . . .

Soil: Fertile and well-drained with plenty of compost topping.

Position: Full sun.

Water: Regularly especially during hot weather.

Top tips: Grow at the right time of year for your location as melons can be a bit sensitive growing out of their temperature range which is soil temps of 21°C and 35°C (70°F and 95°F). They like it hot but not too humid. Trellis or support fruit in a sling, if possible, to keep fruit and vines off the ground and save space. Mulch with straw under the fruit to avoid damp rot due to fruit touching the ground.

Uses: Use in savoury or sweet dishes, pairs well with thinly sliced meats. Use to make sweet or savoury sauces. Best eaten as is and aggressively with juice running down your face. Rockmelon also pairs well with vanilla ice cream and prosciutto (not in the same dish though).

Nutritional power status: Refreshing for your tastebuds and your hydration levels. Bonus of phytonutrients help to prevent chronic disease development.

17. Mulberry

Easy – Round and round we go

Why?: Just like watermelon I have cherished childhood memories of eating my fill of mulberries from trees that were scattered around our neighbourhood. When in season, my friends and I would seek them out in laneways and vacant lots. Years later and going full circle, I've had much joy and satisfaction watching our two young boys growing up with a mulberry tree and doing the same thing.

Soil: Fertile and well-drained but can be very forgiving in a wide range of soils if there isn't waterlogging.

Position: Full sun.

Water: When the tree is young it's good to water regularly until established. Water more during fruiting to help develop big plump berries.

Top tips: Grows very easily from cuttings from a mature plant so if you see a good tree growing somewhere sneak a little cutting and grow your own for free. Mulberry trees can get big and beautiful but pruning them back keeps the tree more manageable and easier to pick the fruit.

Uses: Sauces, jams, or simply eaten fresh from the tree. They can be dried and then taste sweeter. Add to dishes or breakfast cereal. They don't keep well or for long periods in storage like other berries, therefore you will rarely see them on supermarket shelves.

Nutritional power status: Like other berries, they can support mental clarity and may help you to fight off cognitive decline to keep a sharp mind.

Morus spp.
**Silkworm mulberry,
or Common mulberry**

Vaccinium spp.
Varieties: Biloxi, Kisses, Sharpblue

18. Blueberry

Medium – Won't readily grow everywhere

Why?: Blueberries are one of the top natural sources of antioxidants, which protect cells from damage. This is a big-time superfood but I worry how safe commercial blueberries are to eat. Do the pesticides used in growing them counteract their natural goodness?

Soil: Compost rich, well-drained, slightly acidic soil. To provide a more acidic environment you can add some azalea potting mix in the planting hole or container which has a pH of around 5.5 and should help the blueberry plants grow better.

Position: Full sun.

Water: Regular watering every 2–3 days if newly planted until they establish. In hot climates water more in the warmer months of the year as blueberries are not an original warm climate plant.

Top tips: Be patient as it takes a few growing seasons to cultivate a good crop. Blueberries make a great hedging plant and grow well with or around camelias or azaleas for their acidic soil. Protect from birds and insects as the fruit is highly desirable! Use shiny tape tied to a pole to deter the birds or net the plants completely.

Uses: Eaten straight from the tree or freeze if you have a tonne of berries. Make blueberry jam or paste or use in smoothies or juices. Blueberries are a great baking fruit as they are robust and add a beautiful blue colour – one of the few true, blue fruits that release their colour when cooked.

Nutritional power status: They are little powerpacks of nutrition. High in vitamin C and antioxidants, fighting free radicals associated with disease and ageing. They are also a wonderful source of fibre. There have been studies that link their nutritional components to lowering blood pressure.

19. Passionfruit

Easy – If you are passionate about it

Why?: You'll pay $1.50 per passionfruit at the supermarket these days when half the forests in Australia are riddled with passionfruit vines and fruit you can harvest for free. In fact, in parts of Australia passionfruit is considered an invasive weed by environmental groups. That's one tasty weed! So if it can grow that easily why on earth are we buying it from the supermarket? Grow it at home.

Soil: Rich, well-drained soil.

Position: Full sun in a garden bed or a pot with a strong trellis to climb.

Water: Water while establishing the plant and keep soil moist, then deep water once or twice per week in warm weather.

Top tips: Some passionfruit vines from the nursery (especially the black or purple varieties) are grafted on a stronger rootstock. This helps to prevent the vine from getting common root diseases that can kill the vine. But the rootstock can be aggressive and start growing from below the graft eventually taking over the whole plant which is bad because the rootstock may not fruit at all or produce an inferior fruit. So make sure to remove any new shoots from below the graft. One of the best varieties is the yellow Panama Gold which grows true to type from seed, tastes excellent and has good disease resistance so doesn't require grafting onto rootstock.

Uses: Eat the pulp fresh, use in cakes or icing, drinks and fruit salads. They add a unique taste to smoothies, protein shakes and breakfast 'wellness' bowls.

Nutritional power status: Good source of fibre and low in calories, a source of vitamin C and the B group vitamin niacin.

Passiflora edulis
**Varieties: Panama Gold,
Black, Banana**

Malus domestica
Varieties include: Gala, Pink Lady, Red Delicious – and many more!

20. Apple

Easy – An apple a day makes seven in a week

Why?: Apples were one of the first fruit trees I planted when we moved to our acreage back in 2006 because we were *always* buying them. We had young children at the time and apple is easy to cut up for the lunchbox. I'd always wanted to have a go at making apple cider vinegar from our own organic source too.

Soil: Well-drained, nutrient-rich soil. If your soil is clay based add some gypsum and mix through well to break the clay down. However, apple trees can also withstand clay soil.

Position: Full sun.

Water: Apples are surprisingly drought hardy and fast growing too. You'll need to water them regularly to begin with but the tree establishes quickly, and our mature trees survive on rainfall only.

Top tips: Birds (particularly parrots) love apples so netting after flowering is essential. Fruit fly loves them too so make sure the net is fine if you have fruit fly in your area. Apples come from cooler regions so in warmer climates you will need to grow a low chill variety such as Tropical Anna or Golden Dorsett.

Uses: One of the most loved and versatile 'lunch box hardy' fruits. Eat straight from the tree or use in a juice, heat and serve with a sprinkle of brown sugar and yoghurt, bake into muffins, fruit cakes or pies, add to savoury casseroles or use as a puree with roast pork.

Nutritional power status: As the saying goes, an apple a day keeps the doctor away. A great low-calorie snack adding fibre and vitamin C to your day.

20 best herbs/spices

Cavepeople might have eaten staples like yams and meats raw in the beginning, but they soon realised these foods tasted even better by adding some local herbs and spices.

It's no surprise to me that some of the poorest cultures in the world have developed many of the most flavoursome dishes using herbs and spices. I think we can learn a lot from that and experiment ourselves by including more of these aromatic and appetising plants.

For example, I'm always experimenting by combining herbs and spices we grow ourselves to find different flavours and matches for the hero of the recipe. Some combinations don't work very well but occasionally my made-up concoctions turn into eyebrow-raising delights!

My advice to anyone struggling with the cost of living is to invest in a few herbs and spices. Grow them in the garden or in containers on a balcony and then use them to flavour cheaper foods like rice, pasta, chuck steak and other low-cost proteins. You can save hundreds of dollars this way and still eat really well every night.

The following herbs and spices are my best 20 but there are literally hundreds more that you could try. These 20 are a good start though . . .

Zingiber officinale
Ginger root

1.   Ginger

Easy – in warm climates, challenging in cooler areas

Why?: The intense aroma and flavour of this herb/spice that grows so easily in a garden bed is truly incredible. It's a wonder of nature and one of my favourite plants to grow.

Soil: Good, rich and loose compost-infused, fertile soil.

Position: Full sun to part-shade.

Water: Often but don't let soil become waterlogged or boggy (anaerobic).

Top tips: Can be grown in containers on patios, decks or balconies. If you harvest early when young, the flesh is not as pungent as when mature so it's better for eating uncooked in salads or drinks. When mature the stalk dies back and adds to the soil. It's best to harvest it all when this happens and then replant leftovers in spring rather than leaving some rhizomes in the soil. Otherwise, the roots left behind over the winter might rot away.

Uses: Shave into salads, sauté, add to stir-fries, use in teas and juices or candies as a snack. Apparently, it's good for motion sickness (but doesn't help with my seasickness).

Nutritional power status: Research shows ginger may reduce the risk of colorectal cancer and aid in managing nausea and hypertension.

2.   Turmeric

Easy – A wonder health supplement you can grow at home

Why?: Years ago I had no idea a spice as exotic as turmeric was so easy to grow. I knew a bit about it because I've always loved Indian and Asian food, so I'd buy powdered turmeric. Imagine my delight when I realised that it was as easy if not easier to grow than ginger!

Soil: Good, rich, fertile soil is best but will grow practically anywhere in anything.

Position: Full sun to part-shade.

Water: It likes regular water.

Top tips: Can be dried and ground into a powder which will keep for months. Remember to wear protective gloves when washing, peeling or chopping as turmeric stains everything. Return mature pieces to the soil for more growth. Unlike ginger, turmeric pieces are unlikely to rot in the ground over the dormant season so you can harvest as needed and leave the rest to grow again with vigour in spring.

Uses: Add to hot water to make turmeric tea, or add to your latte. Your tongue will turn bright yellow for a while afterwards! Use in savoury baking such as scones or savoury muffins, and in curries and rice dishes to add a vibrant yellow colour.

Nutritional power status: With its strong anti-inflammatory properties, turmeric can help manage arthritis, anxiety and high cholesterol. When paired with black pepper, turmeric is absorbed up to 2000 per cent more in the body!

Curcuma longa
Varieties: Orange, Cape York (Australian Native), Black

Allium sativum

Varieties: Bulb garlic, Hard neck, Soft neck

3. Garlic

Hard – for hot climates (but easy in some areas)

Why?: I love garlic and we eat a lot of it so of course I would want to grow it. Unfortunately, I'm still trying to master how to do so consistently. Some years I grow a tonne of decent bulbs but they're often small. Over the past few years, I've been experimenting by selecting bulbs that have grown well in our garden in the hope I can develop a plant that has adapted to our conditions. I'm confident I got this!

Soil: Free-draining, rich, fertile and compost-enriched soil.

Position: Full sun.

Water: Garlic does like regular watering to grow best but don't waterlog it.

Top tips: Choose a variety suitable for your location. In warmer subtropical areas try soft neck varieties like Glenlarge. Those who live in cool or cold areas are really spoilt for choice with numerous varieties to try.

Uses: Use to flavour dishes instead of salt, add to basil to make pesto, crushed into oil or butter as a sauce. Did someone say garlic bread? How about garlic prawns – yum!

Nutritional power status: Helps keep the arteries clean and controls blood pressure – plus, it's a great addition to make meals extra tasty.

4.   Chilli (both mild and hot)

Easy – in warmer climates

Why?: Chillies are one of the most influential spices on earth. I developed my taste for medium to hot chillies during my military career travelling around the world. And while I don't like super-hot dishes per se, I have grown to enjoy a decent heat with that mouthwatering chilli tang in a curry or condiment. And, technically, chillies could also be a fruitable.

Soil: Good, fertile, free-draining soil.

Position: Full sun.

Water: As necessary for nice plump chillies.

Top tips: Grow for the flavour not the heat. A great decorative plant in containers especially when fruiting. Be careful with hot chillies if you suffer from heartburn or reflux. Chillies contain capsaicin, which slows down digestion and causes food to sit in the stomach longer. The longer food is in the stomach, the greater the risk of heartburn. Spicy food can irritate the oesophagus too, which will worsen your heartburn.

Uses: Mild chillies add a burst of flavour to savoury dishes. Try adding to sweets that have a chocolate or dried fruit base to tantalise your taste buds. Add to vegetable juices or even a weekend cocktail or mocktail such as chilli margarita or Bloody Mary. Finely chop medium to hot chillies and mix well with equal parts vinegar and olive oil with a generous sprinkling of salt and use as a side condiment.

Nutritional power status: Add a bit of spice to your life and you'll reduce your risk of heart-related diseases. Chillies are very high in antioxidant carotenoids, which provide numerous health benefits.

Capsicum spp.
Chilli pepper, Hot pepper

Coriandrum sativum
Leaves = cilantro
Seeds = coriander

5.   Coriander/cilantro

Easy – at the right time of the year

Why?: Some people hate the taste and smell of coriander. To be honest, I was on the fence until I met my wife who loves it and is a big fan of Thai cuisine in which it's used a lot. Over the years, I've grown to love it also and so I grow it. I also dry and grind the seeds into a powder to season and preserve my biltong. I make this tasty South African snack by slow-drying rump steaks.

Soil: Free-draining and not overly fertile soil.

Position: Full sun to part-shade.

Water: Regularly for good-sized plump leaves.

Top tips: I call coriander a 'Goldilocks' plant, in other words, it doesn't like conditions that are too hot or too cold . . . they have to be just right. Coriander 'bolts' to seed very quickly but it will grow well at the right time of the year depending on your climate. Never let coriander stress from underwatering or it will also go to seed.

Uses: Eat the seeds green as a garnish. Dry them to use as a spice. Grind dried seeds to a powder to use in Asian dishes (or to make biltong). Keep some dried seeds in an airtight container to replant next season.

Nutritional power status: Coriander is sometimes referred to as the plant world's 'C' word; you either love it or hate it! It's filled with vitamin K, which can help with bone health and healthy blood clotting.

6. Basil

Easy – Harvest regularly for tender regrowth

Why?: I have European ancestry! And you can't go to the Mediterranean without eating basil. One of the simplest and tastiest meals you can make is pesto alla genovese (or pasta with pesto made in the Genova style). Basil is the key ingredient. Combining basil with mozzarella cheese and my favourite food, tomatoes, is a match made in heaven.

Soil: Very forgiving and will grow in just about all soils but some effort will ensure nice plump leaves rather than leggy stringy ones.

Position: Full sun to part-shade.

Water: Regularly and prune or cut back often to encourage new growth.

Top tips: If it is taking over your garden bed, cut right back as it is a hardy, fast-growing plant. Use the leaves to make pesto sauce that will keep in the fridge for months. If you have a large amount, give jars of pesto away as edible gifts.

Uses: Use as a sauce or the fresh leaves on pizza, in pasta, as garnishes for egg or pancake dishes, paired with tomato and mozzarella for a snack or cooked into bread, scones or muffins.

Nutritional power status: Basil may help to lower your bad cholesterol, while also strengthening your immune system. It may even help reduce stress and anxiety.

Ocimum basilicum
Sweet basil, Thai basil *(Ocimum basilicum var. thyrsiflora)*

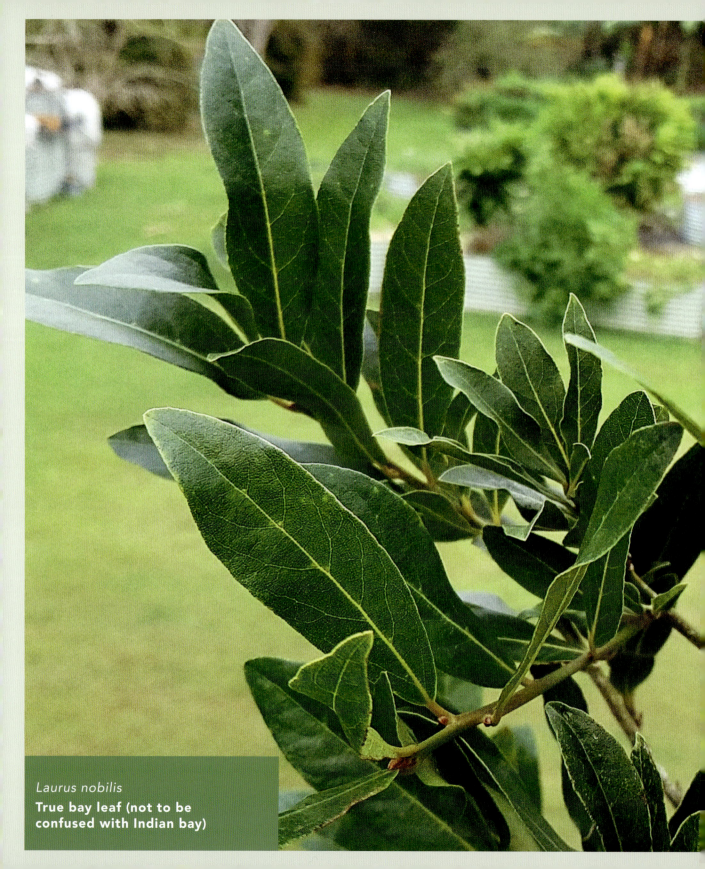

Laurus nobilis
True bay leaf (not to be confused with Indian bay)

7.   Bay leaf

Easy – Slow grower, but worth the wait

Why?: We bought our first bay tree as a small seedling back in 2002 when we were in Victoria. A few years later we moved to Brisbane, so we put the tree in a pot on our trailer and brought it with us. Unfortunately, the plant wasn't covered so when we arrived all the leaves had been stripped away. Anyway, miraculously it survived long enough to be planted in the ground. That bay tree is now 5–6 metres tall and thriving!

Soil: Very forgiving. Prefers free-draining soil. Can be grown as a living herb hedge.

Position: Full sun.

Water: Bay trees are native to the Mediterranean. Water moderately as they don't like wet feet but they also don't like to dry out. The roots are very shallow and frequent watering may be necessary during dry months.

Top tips: Will grow up to 5 metres (15 feet) sometimes bigger in ideal conditions but can always be pruned back. An ornamental tree for your garden that doubles as a lovely aromatic herb for cooking. They are slow growers so be patient. Use the leaves fresh rather than dried for a fuller flavour.

Uses: Throw the leaves into simmering dishes with meat or chicken, or flavour lamb shanks. Add to soups or flavour plain rice. Grind dried leaves into a powder and add to your Italian spice repertoire.

Nutritional power status: While the aroma of bay leaf can make your kitchen smell incredible, it can also help to relieve an upset stomach and support a healthy immune system.

8. Rosemary

Easy – A sprig of rosemary . . .

Why?: Australians traditionally wear sprigs of rosemary as a symbol of remembrance on Anzac Day or Remembrance Day. It's an ancient symbol of fidelity. I think remembering those who sacrificed their lives to keep our way of life should always be cherished. Rosemary also happens to be a wonderful herb to use in cooking, particularly with lamb. Grow it on either side of a pathway so every time you walk past the perfume wafts in the air.

Soil: Well-drained and fertile, however very forgiving and will grow in garden beds, in the ground or hanging pots or window boxes.

Position: Full sun.

Water: Regularly, but is quite drought tolerant.

Top tips: Harvest and cut back regularly to help the plant 'bush out', avoiding a 'leggy or spindly' plant. Regrows easily from small young cuttings. Rosemary can grow large if left unpruned (a metre or 3 feet high and wide) so it can make a nice medium-sized edible hedge.

Uses: Add to your Italian herbs in sauces, soups and stews, spike into meat dishes so the flavour cooks into the meat. Add to breadcrumbs to flavour a crust or make your own croûtons. Grind up with coarse salt using a pestle and mortar and add olive oil. Paint onto pork, chicken or chops.

Nutritional power status: Having antibacterial and antiviral properties, the aroma will not only boost your mental state, but it will protect your immune system too.

Salvia rosmarinus
(Formerly *Rosmarinus officinalis*)

Aloysia citrodora
Lippia citriodora, Lemon beebrush, Vervain citronnelle

9. Lemon verbena

Easy – Makes a nice tea!

Why?: I didn't know much about this herb until I saw it at the nursery and thought I'd give it a go. It's a lovely, scented plant and I'm surprised that I hadn't heard of it for all these years and only started growing it about five years ago. That's gardening though, you never stop discovering new plants or ideas.

Soil: Grows well in basic soils. Likes containers and will withstand long dry periods.

Position: Full sun.

Water: Regularly but is very drought tolerant.

Top tips: A perennial herb that dies back in winter and bursts into life in spring.

Uses: Add to teas, cakes and tarts, sauces, and fish dishes. Adds a refreshing flavour to desserts such as ice cream and fruit dishes, and vinaigrettes. It's an excellent ingredient in herbal teas and has a noticeable and lingering sweet lemon flavour.

Nutritional power status: Exciting evidence shows it can help speed up muscle recovery post-exercise.

10. Thyme

Easy – In fact, super easy

Why?: The easiest way to get more time in your life is to grow it. A Mediterranean herb, it will grow just about anywhere including in shallow rock gardens, in pots, on windowsills and in garden beds.

Soil: Tolerates many soil types. It is a small undergrowth plant that does not need a lot of space.

Position: Full sun. Will grow under other bushes such as tomatoes and can be poked into rocky outcrops.

Water: As required. In very hot conditions water 1–2 times a week. Is quite drought tolerant.

Top tips: Remove leaves from the stalk before cooking as they fall off in the dish leaving behind the annoying stalk. Use fresh from the bush and don't store in moist climates as thyme loses its flavour. Cut back occasionally to encourage new growth.

Uses: Pairs beautifully with lemon and is great for flavouring fish dishes. Add to French soups and sauces. Pair with oregano and marjoram to complement lamb, tomatoes and starchy vegetables.

Nutritional power status: Rich in vitamins, including vitamin B6 which can help to relieve stress, and vitamin A to maintain cell membranes, skin and keep your vision sharp.

Thymus vulgaris
**Varieties: Lemon,
Creeping, Common**

Origanum vulgare
Mediterranean oregano

11. Oregano

Easy – will thrive on neglect

Why?: Another native Mediterranean herb that is a must-grow at home. Dried and mixed with basil and thyme sprinkled over your homemade bruschetta your family and guests will be herbaceously impressed!

Soil: Easily grows in all soils.

Position: Full sun.

Water: As required however is quite drought tolerant.

Top tips: If you only have a small area, grow with other Italian herbs in the one pot or container so you can harvest everything for the meal at the one time. The flavour intensifies when grown in a full day of sunshine so move the pot to the required spot for the day before harvesting. When the plant becomes sparse and leggy give it a very short haircut and then cover with compost and it will grow back bushy and succulent.

Uses: Pairs brilliantly with tomatoes and olive oil. Use when making your own bread or damper. Sprinkle over bruschetta, in pastas or on pizzas.

Nutritional power status: High antioxidant activity helps to prevent cell damage and contains a rich source of vitamin K for the protein osteocalcin to support healthy bones.

12. Lemongrass

Easy as growing grass

Why?: If you are into Asian cooking, lemongrass is essential. It does have a strong growing habit that will block out other grasses or plants but it's not invasive. Lemongrass is the kind of plant you buy once and then divide forever to make many more.

Soil: Grows in a wide range of soils but prefers a rich, free-draining soil.

Position: Full sun as lemongrass loves warmth.

Water: As required and to establish but can be tolerant of long dry periods.

Top tips: Use as a grass hedge to 'crowd out' other weeds and grasses. Very attractive planted along fence lines and easy to propagate through the stems, pulling and dividing and replanting. Easier to grow by dividing than from seed but you can do that also if you want by harvesting the mature seeds and sowing in punnets.

Uses: Lemongrass tea, rolled up, boiled and strained. Use in dishes such as chicken or pork larb, fish curry, dipping sauces, finely sprinkled on meat skewers, tossed through glass noodle salads and vegetable dishes.

Nutritional power status: Disease-fighting properties make lemongrass more than just a flavoursome addition to your cooking. It also helps to fight inflammation of gums and can prevent cavities.

Cymbopogon citratus
West Indian Lemongrass

Alpinia galanga
Thai ginger

13. Galangal

Easy – Not well known but well used

Why?: Galangal is related to ginger and is often a 'hidden' herb in Thai cooking. Most people can't pick it but will notice a flavour drop if it's missing from curries, hot and sour soup, and Thai beef stew. But galangal is not just used by the Thais – there's Malaysian/Indonesian beef rendang, Chinese chicken and thousands more recipes.

Soil: Free-draining soil.

Position: Grows just about anywhere including part-shade and even in boggy areas.

Water: Regularly – it does like a good drink.

Top tips: Grows in a clump like ginger but two or three times the size in height so allow plenty of room. Don't plant it in the middle of your vegetable patch like I initially did because it will quickly become a feature. I plant it like a fruit tree. The rhizome or root system is used for cooking and at around $30–40 per kilogram in supermarkets it's worth growing yourself! If you like Asian cooking, it's an ingredient that should not be excluded and can be used dried, powdered and fresh.

Uses: A more citrusy tasting ginger. Used in Asian and Indian cooking, its flavour is peppery and spicy with a zesty bite and a hint of pine.

Nutritional power status: Galangal may benefit male reproductive health and has excellent anti-inflammatory properties – making this spice great for preventing chronic disease.

14. Dill

Easy – Don't be a dill and not grow this!

Why?: I love fermented dill pickles so growing dill for me is essential. But I'm also a keen fisherman and dill goes well with fish. One thing I have noticed when preserving dill is normal drying or dehydrating is not as flavourful as freeze-drying.

Soil: Grows well in a range of soils and will grow in a pot or garden bed.

Position: Full sun to part-shade.

Water: As required, just don't let it dry out.

Top tips: Always let dill plants go to seed as they attract good insects and bees to your garden. When seeds grow into what looks like a carrot bloom, save the seeds to replant and/or grind into a powder. Mix the dill powder (1 teaspoon) with mayonnaise (1 tablespoon) to use as a sauce condiment with white fish or on fish burgers (one of my personal favourites).

Uses: Pairs well with pickled cucumber and potatoes, boiled or cold potato salad. Great for pickling other vegetables. Will last well when dried.

Nutritional power status: Don't be a dill – add this delicious herb to elevate your meal and you'll help your body fend off inflammation.

Anethum graveolens
Dill weed, Dill seed, Lao coriander

Allium schoenoprasum, **Onion chives**
Allium tuberosum, **Garlic Chives**

15. Chives

Easy — Looks like grass, smells and tastes like garlic or onion

Why?: Chives just grow in the garden without any fuss or admiration until you need them and then you're happy they're there. Garlic chives have a surprisingly strong garlic smell and taste making them an excellent substitute for real garlic. Onion chives must be in a potato salad or all hell will break loose in our kitchen!

Soil: Very hardy in most soil types or potting mixes.

Position: Full sun. Grows well in containers or as garden bed edging.

Water: Regular watering is not essential as chives can survive on small amounts of natural rainwater.

Top tips: Grows all year round in most climates. Try different varieties such as garlic chives or society garlic (not really chives but similar) if you don't tolerate traditional garlic (such as people with IBS). Divide plant regularly to grow more. Easily grows from seed.

Uses: Use in sauces, soups, savoury dips for snacks or entertaining. Add to mashed or baked potatoes, fish or seafood dishes and egg meals such as scrambled eggs or an omelette. The delicate flavour of chives can be minimised by heat so add to dishes at the last minute for better flavour.

Nutritional power status: Although mainly eaten in small quantities, chives contribute to fighting inflammation and can introduce healthy bacteria to support a more diverse microbiome.

16. Sage

Medium – Can be a bit difficult in humid climates

Why?: It helps flavour stuffing for roast chicken and everything else is a bonus after that. I've never been able to successfully grow sage as a perennial because it tends to suffer in our hot and humid summers. It does grow well through the cooler months though.

Soil: Relatively fertile and well-drained.

Position: A bit of shade is good in hot climates and it grows well when mixed with other herbed plants.

Water: Keep the moisture up but don't overwater.

Top tips: Use other herbs planted around sage to protect it. It's been used by people to cleanse their homes in a process known as 'smudging' – burning the tips of the leaves to add aroma to the home.

Uses: Adds a beautiful flavour to most proteins. Make crispy sage leaves to garnish by frying in butter. Can also be used to add an intense flavour to sauces, butters and spreads, marinades for meat, pastries, and breads. Add fresh sage leaves to cocktails and teas for an instant herbal flavour.

Nutritional power status: Strong antioxidant and antibacterial properties can help keep your immune system intact.

Salvia officinalis
Common sage, Garden sage, True sage

Foeniculum vulgare
Florence fennel (bulb)

17. Fennel

Easy – And underrated . . .

Why?: There are a tonne of uses for fennel both as a bulb, foliage and also for the seeds. Could it be a 'Veggieherb'? I'm very tempted to coin another word, but I'm calling it a herb/spice.

Soil: Free-draining and fertile soil for best bulbs/bases.

Position: Full sun but does okay in part-shade.

Water: Regularly to keep juicy and plump as water stress can make the plant fibrous.

Top tips: Try the purple variety instead of the regular green one; it is unusual and visually spectacular in the garden (and in dishes).

Uses: The distinct aniseed flavour adds zest to salads and vegetable dishes and cooks well into stir-fries and casseroles. Chop or shave the bulb into salads and use the fennel leaves like dill.

Nutritional power status: Great source of essential fatty acids, magnesium and potassium – which all help to support brain, muscle and nerve function and keep your heart pumping.

18. Parsley

Easy — It's not just a garnish for bistro meals!

Why?: Tabouli salad is pure genius and I can't help think about this brilliant but simple Mediterranean dish whenever I see parsley growing in the garden. Did you know that curly or flat leaf parsley are from the same family as carrots? That's why carrot tops have that similar parsley smell and taste.

Soil: Grows in just about any soil – extremely hardy.

Position: Part-shade or full sun.

Water: Regularly for strong stems and plump leaves.

Top tips: Experiment with both the Italian flat leaf variety, which has a milder flavour, and the curly variety, which has a stronger taste. Flat leaf parsley also propagates and goes to seed more easily than the curly, and this is handy for backyard growers wanting to replenish the older more woody plants with younger ones.

Uses: Use curly on pasta and flat leaf in salads. Both parsley types are great on grilled vegetables, roasted potatoes, stews, soups or hot or cold grain dishes like couscous, quinoa or tabouli.

Nutritional power status: Rich in vitamin K which benefits blood clotting and building bone, and is also high in zeaxanthin, which can prevent age-related macular degeneration.

Petroselinum crispum
Flat-leaf and Curly-leaf varieties

Mentha spp.
Includes Spearmint, Peppermint

19. Mint

Easy – Mojitos here I come!

Why?: For mojitos, naturally! And homemade mint sauce ain't bad either and goes superbly with roast lamb. I've had the same mint crop growing in a round raised garden bed for almost two decades now and it continues to thrive.

Soil: Free-draining, rich soil is best.

Position: Very resilient and grows via runners that extend swiftly through the soil. Will find its own sunny/part-shade position to grow like wildfire.

Water: Regularly for plump leaves.

Top tips: Isolate it, preferably in a pot or contained raised bed as it is invasive and will take over the rest of the garden – and be difficult to completely remove.

Uses: Refreshing cold drinks, use with Angostura bitters as a low-alcohol alternative. Use in Vietnamese dishes such as rice paper rolls and noodle soups. Add to cooked peas to supercharge their flavour.

Nutritional power status: Not only is mint great for giving a refreshing taste, it's also an easy addition to a meal (or drink) to contribute to your gut bacteria diversity. Grab a sprig from your garden before you leave home for the day . . . it's better than chewing gum for freshening your breath.

20. Curry leaf tree

Easy – Sometimes too easy

Why?: We make Indian curries all the time at home and many authentic Indian dishes require curry leaves from their native curry leaf tree.

Soil: Normal, free-draining soil.

Water: Regularly however is drought hardy.

Top tips: Can grow to around 4 metres (12 feet) high so give it space or prune regularly for more leaves. The berries can be eaten (not the seeds), but it's the leaves that are typically used in cooking. Fresh berries will strike easily, and this can lead to seedlings popping up around the tree and the yard. For this reason, in some Australian states the curry leaf tree is considered an invasive weed.

Uses: Adding that missing ingredient to curries and flavoursome stir-fries. Lovely lemongrass flavour to boost sauces, dressings and make savoury jams and chutneys.

Nutritional power status: Studies show that inhaling the linalool in curry leaves can reduce anxiety, stress and help with depression. Apply 2 to 3 drops of curry leaf oil on your pillow before you go to sleep. The scent has been shown to calm your mind and body.

Murraya koenigii
Curry leaf plant, Curry bush

20 best unusual food crops

I selected these 20 'unusual' food crops because many of them are not very well known but they're some of my favourites to grow at home. You might find some of them in the supermarket, but they're often very expensive or hidden away on a shelf at the back of the fresh food area slowly decaying because few people care for them.

I'm always on the lookout for unusual food crops because (as you probably know) growing food fascinates me, it's my passion, and I've stumbled on fruits and vegetables I'd never heard of that turned into some of my favourite things to eat.

If you hunt for new foods to try at the local supermarket, you're searching in the wrong area. It's like attending a political rally and expecting to hear new ideas on how to fix the economy – it's just not going to happen. You need to go to farmers' markets, community markets, plant shows, local food growers' get-togethers, off-the-beaten-track nurseries or search online for independent seed and food plant sellers.

Anyway, here are my best 20 unusual food crops to grow. I hope you like them.

1.   Rosella

Easy – What a lovely fella

Why?: I grew up seeing rosella jam sold at every school fete I ever went to but it wasn't until I was an adult that I realised what a rosella plant actually looks like. Rosella isn't native to Australia but it's such an iconic and tasty jam typically made at home that you'd think it was.

Soil: Free-draining, fertile soil for good calyx (the outer part of the flower) development.

Position: Full sun and prefers the warm weather through summer.

Water: Rosella likes a lot of water and this helps plump up those jam-making calyxes.

Top tips: This type of edible hibiscus grows to around 1 metre tall and 1 metre (3 feet by 3 feet) wide so be careful how close you plant next to others in your garden bed. They only last one season, quickly dying back as the weather cools.

Uses: Apart from jams, the calyx makes refreshing cordials and teas; you can add a sweetener such as sugar to improve the taste as they are not sweet themselves. The seeds contain pectin which is the gelling agent that helps set tarts, jams and pies.

Nutritional power status: Whether used in juices, tea or jams, rosella is an excellent source of anthocyanins which can help to lower bad cholesterol in the blood, keeping your heart in good nick.

Hibiscus sabdariffa
Also known as Jamaican sorrel

Lagenaria siceraria
Also called Bottle gourd

2.   New Guinea bean

Medium — takes up a lot of space

Why?: Another name is Italian gourd and that's the main reason why I decided to give this large zucchini-like vegetable a go. I thought to myself, 'If the Italians like it then it's probably good.' And it is!

Soil: Fertile, free-draining soil.

Position: Full sun and requires a strong trellis or 'gourd tunnel' to grow over.

Water: Plenty of water. It can produce huge fruit that require large amounts of water to develop properly.

Top tips: Harvest when young and before they get too big for a tastier and less woody fruit. The fruit can be up to a metre (3 feet) long but the best size for eating is around 30 centimetres (12 inches) or so. Fruit fly can strike the young fruits causing them to drop if not developed properly so some type of bagging or netting is helpful.

Uses: Eat and use like you would a zucchini. But unlike zucchini, the benefit of the New Guinea bean is that it easily grows through the hot and humid summers and won't die off from fungal diseases like powdery mildew.

Nutritional power status: Sometimes known as Cucuzza squash, this green vegetable has a high potassium content which can help regulate blood pressure.

3.   Taro

Easy – Survival crop

Why?: Taro can grow in normal soil conditions, boggy ground or even in a pond of water. These growing qualities are rare and with all the rain we've had over the last decade or so, taro is a no-fuss crop to plant and forget. In addition, the high-energy carbohydrate makes it a great survival crop should things turn to custard in the world.

Soil: An extremely wide range of soil conditions.

Position: Full sun is best but it can grow under trees in dappled shade.

Water: Likes lots of water so do water often.

Top tips: Grows well in any large container, inground or in raised beds. Start with a small pup (seedlings develop from a parent plant), tuber or stem cutting planted directly where they are to grow. Keep well-watered and the plant should grow well through the spring and summer. Growth slows considerably in the cooler months.

Uses: Can substitute for potatoes and is slightly healthier with more resistant starch than many other starchy vegetables. Use as a mash, make into wedges, patties, burgers or fish cakes.

Nutritional power status: A great source of fibre which can help with digestion, and it's a great non-animal source of zinc to promote healthy metabolism and immune systems.

Colocasia esculenta
**Gabi (Philippines),
Dasheen (Caribbean)**

Helianthus tuberosus
Also called Sunchoke

4. Jerusalem artichoke

Easy – but can be hard on the guts for some

Why?: 'Fartychoke' is its nickname and that's all I needed to know to want to give this veggie a go in our garden! Yes, I can say from personal experience that it's true about giving you wind but you can train your stomach to process Jerusalem artichokes better by eating them more regularly. I also found out through experimentation that they ferment well and become a delicious pickle (with fewer gas problems).

Soil: Crumbly, loose soil so the tubers can expand and grow well.

Position: Full sun.

Water: Keep them well-watered, especially during hot periods.

Top tips: This beautiful ornamental plant needs to grow in its own container or container area as it can be invasive. The edible tubers grow under the soil and the edible flowers aboveground are used by insects to pollinate other flowers. Once dug up, the tubers will go soft. The best way to keep them crunchy is to pickle or ferment them. The reason they're called 'fartychokes' is because of their high fibre content.

Uses: Has a creamy nutty flavour therefore a brilliant addition to baked pies, baked vegetable flans, or use marinated on pizzas, in pasta or on an antipasto platter.

Nutritional power status: The high inulin content is a game-changer for promoting a healthy gut microbiome. If you're passing more wind than normal . . . then rejoice! Your gut numbers are healthy.

5. Egyptian spinach

Easy – A great emergency green

Why?: When we moved to our acreage outside Brisbane and started growing food I quickly realised that there was an ever-widening gap of three to four months per year through our hot summer when we couldn't grow traditional salad crops like lettuce and English spinach. This was when I started experimenting with substitutes and discovered Egyptian spinach.

Soil: Good, fertile and free-draining soil.

Position: Full sun but will grow in part-shade.

Water: Regularly for plump leaves.

Top tips: Egyptian spinach will grow when the English spinach season is over – making it a great replacement.

Likes to grow in the warmer summer months and will readily seed with these seeds remaining dormant until the next summer when they germinate. These seedlings can be left to grow in place or transplanted to other beds. It's a very easy and economical plant to grow.

Uses: Use as you would the English variety but the Egyptian flavour is a lot milder. Perfect for salads, quiches and stir-fries.

Nutritional power status: A powerhouse vegetable! Containing more than 30 vitamins and minerals, a good source of magnesium, it can help protect against osteoporosis and improve circulation to boost energy levels.

Corchorus olitorius
Molokhia, Jute mallow

Ipomoea aquatica
Water spinach, Ong choy

6.   Kangkong

Easy – Can do!

Why?: Also known as water spinach, kangkong is the King Kong of the garden. Not really, but I wanted to fit that in somehow just for fun. Seriously, go to any Asian grocer and you'll find kangkong. It's enjoyed for its mild and pleasant flavour, nutrient-dense qualities, and easiness to grow in hot humid conditions. I guess it's king for those reasons.

Soil: Nutrient-rich, evenly wet and moist soil.

Position: Full sun.

Water: Likes lots of water, can even grow in a pond.

Top tips: Kangkong will grow well in tough, hot and humid conditions, even when traditional greens won't, making it a great alternative in the off season for warm climates. Harvest regularly to encourage new growth.

Uses: Use the whole plant. The stem and leaves can be used as an Asian vegetable in stir-fries and in savoury soups.

Nutritional power status: High in iron, potassium and calcium – great for those who need to boost their iron levels and want a sweeter style of spinach alternative.

7. Globe artichoke

Medium – Spectacular flowers

Why?: This is another Italian favourite. Because I grew up eating artichokes, the natural thing for me was to grow them and learn how to use this unusual thistle from the Asteraceae family (same as a daisy but looks nothing like one).

Soil: Free-draining soil.

Position: Full sun.

Water: Plenty in hot climates, easier to grow in cooler climates.

Top tips: It can take a while to get a good crop – sometimes up to two seasons – however be patient as it's well worth the journey. Globe artichokes are usually grown as a perennial for about 2–3 years and then the crop needs to be refreshed by growing again from pups produced at the base of the parent plant or via seed germination from mature flowers.

Uses: A bit of effort to prepare but totally worth it. Has a well-known and unique taste and is popular when preserved in oil for antipasto plates. Harvest the buds young before they open. Shave the bud back and boil in water. The heart of the bud and sometimes the base of the larger petals are eaten often dipped in butter with a squeeze of lemon. Or cook and eat the leaves with your choice of dipping sauce, toss on the grill with lemon and oil or add to your favourite cheese platter instead of crackers.

Nutritional power status:
One of the richest sources of polyphenols among vegetables, making it a great way to protect your liver and lower cholesterol.

Cynara cardunculus var. scolymus
Green Globe, Purple Globe

Hylocereus undatus

Pitaya. Varieties: Yellow spiny, Red, Pink, Purple

8.   Dragon fruit

Medium – An amazing cactus

Why?: Some people think dragon fruit tastes bland but I strongly disagree! I think the reason why people feel this way is due to eating one bought at the supermarket. I've tried several supermarket dragon fruits and they have been less flavoursome than the ones we grow at home. This is probably due to the time between harvesting and eating. The longer the time, the more the flavour retreats so best to eat them soon after picking.

Soil: Very free-draining, almost sandy soil.

Position: Full sun although can grow in part-shade as it's typically a rainforest cactus.

Water: Even though it's a cactus and can survive long dry periods, it needs water regularly and often for best results. The more water the better especially when flowering and setting fruit.

Top tips: In cold climates, grow in containers and move into a green house in winter or cooler months. Needs a post or strong trellis structure especially when fruits appear as they are heavy plants. Hand pollinate if you find fruit is not setting.

Uses: Looks spectacular in fresh fruit salads; try the white, pink or red varieties. Use in smoothies, cut into ball shapes and add to salads, blend into ice cream, thread onto skewers with other melons.

Nutritional power status: Rich in essential amino acids, fibre and vitamin C, regular consumption has been linked with effective wound healing and stronger immunity. Eating a lot may turn your pee pink!

9. Finger lime

Easy – Known as citrus caviar

Why?: This is one of Australia's best native edibles so I just had to grow several varieties for myself. I like to squeeze the little caviar-like spheres out of the elongated fruit and eat it like that. You need to be careful of the spikes or thorns that cover the tree, but it does make a good edible front hedge against the fence deterring unwanted intruders!

Soil: Well-drained and will grow in normal garden soils with a citrus fertiliser.

Position: Full sun although will grow reasonably well in part-shade as historically it's a rainforest plant.

Water: Requires a lot of water especially when fruiting.

Top tips: This medium-sized citrus bush grows well in its own pot. Grows easily from seed, but will fruit more quickly from a cutting grafted onto citrus rootstock.

Uses: Very popular in modern cuisines where the tiny, flavoursome sphere-like pulp is used as a garnish or addition to a dish. Enjoy them in a sweet fruit salad or a savoury salad or replace citrus juice in a vinaigrette with finger lime pulp for a burst of flavour on seafood or as an oyster garnish. Finger lime pulp can be used in a citrus curd or mixed into a cheesecake. Or try finger limes in biscuits, ice cream or on top of desserts with whipped cream.

Nutritional power status: With their citrusy bursts of flavour, they make a fancy garnish and give you a great hit of vitamin C for a stronger immune system. They may also aid in collagen formation for smoother skin.

Citrus australasica
Native Australian citrus

Carya illinoinensis
Varieties: Shoshonii, Cherokee

10. Pecan

Easy – Astoundingly fast-growing tree

Why?: Pecan pie or pecan brittle are two desserts that really hit the spot for me. Before getting our first pecan nut tree I thought I knew how big they can get but my plan was to never let it get so big it couldn't be managed. It took about four years to fruit and the tree was around 4 metres (12 feet) high. Our harvest was exceptional and pecan pies were on the menu! I stuck to my plan and pruned it to around head height. That was a few years ago and already it's towering over the landscape. Unfortunately, as time gets away from you so did this tree and although it fruits profusely, we are getting no nuts at all because the cockatoos take advantage of the height and graze away with impunity. Yes, I can't wait to give it the chop.

Soil: Will grow in most conditions but not in boggy ground.

Position: Full sun.

Water: Moderate depending on your location.

Top tips: Keep the tree small and manageable! Large trees attract birds such as cockatoos and they will destroy your harvest before you get to it. The nuts develop quickly and within a few months they are ready to eat. It's a handsome tree with soft leaves and is worth planting for ornamental reasons alone.

Uses: Eat straight from the shell or use in pies and tarts. Bake into a pecan brittle, crush as a savoury crust on beef or lamb, or use in a stuffing recipe for roast poultry. Pecan meal is a great gluten-free alternative to flour.

Nutritional power status: Go nuts for pecans – they contain healthy fats and vitamin E to help fight off a cold or flu, and have been shown to reduce the risk of cardio metabolic disease.

11. Barbados cherry

Medium – Nothing like a real cherry

Why?: Who doesn't like cherries? Our Barbados 'cherry' or Acerola tree was one of the first fruit trees I planted on our property because while we can grow lots of different fruit trees in our climate, real cherries aren't one of them – it's just too hot. So the compromise is to grow something that does like warm growing conditions but also tastes like a cherry. Well, the Barbados cherry grows well in warm climates but it's nothing like a real cherry! But it's still a good fruit and totally worth growing.

Soil: Most soils but hates wet feet/boggy ground.

Position: Full sun.

Water: Regularly when in flower so the fruits develop well.

Top tips: Good ornamental value as well as an edible plant. Keep an eye out for citrus borer as it will attack this tree so if you see sudden dieback of branches find the borer and kill it quickly.

Uses: Can be eaten raw but they may be a bit tart for some. The intense red colouring of the fruit makes them perfect for use in jellies and jams; drinks including non-alcoholic and alcoholic beverages such as wine; desserts such as ice creams, tarts and pies; they also make tasty syrups and sauces.

Nutritional power status: Valued for being one of the top fruits for vitamin C, they are also incredibly high in phytonutrients to help reduce risk of cancer and stroke.

Malpighia emarginata
Acerola cherry

Solanum betaceum
Tree tomato

12. Tamarillo

Medium – Also called tree tomato

Why?: When I heard about this 'tree tomato' I was intrigued and had to buy some! Tamarillo is related to tomatoes and they share some obvious commonalities like leaf shape and flowers. Even the fruit has some distant similarities. I first grew them in the ground like any other fruit tree and subsequently found out they hate clay soil and will slowly succumb to root rot and die. Now I grow them in raised beds or containers. The fruit can be a little sour but I really like them and find the taste a bit like a kiwi fruit.

Soil: Very well-drained or it will die.

Position: Full sun to part-shade and if your soil is the slightest bit heavy grow in a container.

Water: Needs regular watering especially during flowering and fruiting; be careful not to let the soil get bogged.

Top tips: From seed they'll fruit within 12 months and grow to 3 metres (10 feet). I like to prune the top to make a smaller, more compact plant. You get around five years from the plant before it dies so try succession planting to ensure a consistent supply.

Uses: Salsa, chutneys and spicy sauces work well. Makes great fruit leather for lunchboxes. Puree the fruit and strain to remove pith, then dry to turn it into a leather-like consistency. This is much healthier than the commercial roll ups for kids (and adults). Remove the skin as it can be bitter and use the flesh in ice cream, gelati or sorbets. Serve cooked tamarillo with poultry or fish.

Nutritional power status: Home to a whole host of vitamins and minerals, tamarillo can ensure healthy eyesight and protect you from infections thanks to its high antioxidant content.

13. Yacon

Easy – Also known as ground apple

Why?: In my search for plants that can be used instead of sugar I followed my nose to yacon, which is a plant that grows a tuber similar in size to a small sweet potato. It's sweeter than a sweet potato though and crunchy like an apple. It can be eaten raw just like an apple and it's pleasant tasting! Can also be juiced and added to cooking or drinks as a healthier option to refined cane sugar.

Soil: Well-drained, fertile soil.

Position: Full sun. Prefers a warm environment and doesn't tolerate frosts.

Water: Water in well when first planted, yacon are thirsty plants especially in hot weather. Well-drained soil will ensure water doesn't sit around the base of the plant and rot the tubers.

Top tips: Give them plenty of space as the root system can grow to 2 metres (6 feet). It will keep expanding if not dug up. Plant in late winter when frost risk is over. Harvest when mature and replant the rhizomes (not the tubers) to grow more plants.

Uses: I think they taste like a nashi pear while others reckon they're a mix of cucumber and apple. Can be used in stir-fries and add great flavour to soups and casseroles. Roast with other tubers or add to juices or desserts.

Nutritional power status: Can be eaten raw providing a rich fibre source. Also low in calories so a healthy snack. They contain a type of resistant carbohydrate called FOS, a prebiotic which feeds up the good bacterial in your gut (and yes, that means more wind)!

Smallanthus sonchifolius
**Peruvian Ground Apple,
Groundpear**

Amorphophallus paeoniifolius
**Suran (India),
Whitespot giant arum**

14. Elephant yam

Easy – Looks like an elephant's foot

Why?: Survival crops interest me and the elephant yam is another one of those food plants that grow easily in my climate, take little effort to maintain and taste really good!

Soil: Good fertile soil not too compacted.

Position: Full sun.

Water: Regular watering.

Top tips: It's a heat-loving plant that dies back in winter and remerges mid-spring through green shoots that turn into tall, taro-like leaves. The plant produces round, flat tubers that look like an elephant's foot (it's quite uncanny) which can be eaten like any yam or sweet potato. Tubers can be dug up and relocated to make new plants and stem cuttings can be taken to propagate also.

Uses: Cut the tubers into chips and fry in oil or bake. The stems of the plant are also edible sautéed like any green vegetable.

Nutritional power status: Rich in energy-boosting carbs and some vegetable protein, it is also loaded with zinc, phosphorous, potassium, vitamin B6, vitamin A and calcium. Providing phenols, alkaloids and flavonoids which is 'science talk' for maintaining proper body functions.

15. Macadamia nuts

Easy – Not as mainstream as you'd expect

Why?: Another Australian native food that is worth growing. Most people love macadamia nuts but you don't see them in many backyards. I think this is due to the laborious nature of collecting the nuts, waiting months for them to dry, and then cracking the impossibly hard shells to get to the tasty kernel inside. It's much easier to buy the pesticide-laden product already cracked, salted and packaged for a premium price at the supermarket! Yes, that's sarcasm . . . Just buy a macadamia nutcracker and grow your own chemical-free. It's not that hard. Well, the nuts are . . .

Soil: Grows in most soils.

Position: Prefers full sun.

Water: Regular watering but once established very drought tolerant like most natives.

Top tips: Pick a variety that supplies plentiful fruit. There is one actually called 'lots of nuts'. Some varieties can be sparse and more susceptible to pests. Varieties with a smaller nut tend to bear more fruit.

Uses: Eat as a snack with fruit or crush to use as a tasty crust on lamb chops or whole fish fillets. Mix into your favourite cereal or roast and blend into a tasty paste for toast.

Nutritional power status: Healthy monounsaturated fats balance out LDL cholesterol with good HDL cholesterol keeping your heart healthy.

Macadamia integrifolia
Queensland Nut Tree, Bush Nut

Lotus tetragonolobus
Winged pea

16. Asparagus pea

Easy – Also called winged bean

Why?: This is another crop that thrives during our hot summers when regular peas and beans don't. It's a bean and they are best eaten when small and young or they quickly become stringy. The taste is a bit like asparagus.

Soil: Good, free-draining and fertile soil.

Position: Full sun and best grown up a trellis.

Water: Regular watering will produce better 'fruits' or pods.

Top tips: Give them their own garden space and trellis as they are aggressive growers and will take over an area. Grow like regular peas about 10 centimetres (4 inches) apart to fit in as many vines as possible to produce more same-sized pods to harvest at the same time.

Uses: Stir-fries, salads, steamed, boiled, but you don't generally shell them.

Nutritional power status: Used as greens, they're an excellent source of fibre, vitamin C and vitamin A. Half a cup of fresh leaves provide 45 milligrams of vitamin C (close to your daily requirement). The tubers are also rich sources of starch, vegetable protein and B-complex vitamins.

17. Jaboticaba

Easy – Cool name

Why?: This tree produces big, round, marble-sized grape-like fruits straight from the trunk. I read about the jaboticaba browsing some fruit tree retailers online and decided to give it a go. Our tree is about five years old and not very developed at 3 metres (10 feet) high so we haven't had a lot of fruit yet. However, the fruit that we have harvested has been fantastic!

Soil: As usual and, like most fruit trees, free drainage is necessary.

Position: Full sun.

Water: Regular watering to establish the tree.

Top tips: So far, our tree has been easy to grow with no noticeable problems. The fruit doesn't seem to attract fruit fly sting. It can grow to 15 metres (45 feet) so we'll be pruning ours to keep it under 5 metres (15 feet).

Uses: A delicious fruit straight off the tree. People use it to make jams and wines.

Nutritional power status: Contains appetite-supressing properties which may have beneficial effects on metabolic diseases. And no prescription needed.

Plinia cauliflora
Guaperu, Brazilian Grape

Cucumis metuliferus
Kiwano, Horned melon

18. African horned cucumber

Medium – Big vine

Why?: This horny devil of a fruit would not have been in the running to make this list a few years ago because I had made my mind up that it was not worth growing. But as luck would have it a self-seeded vine started growing over one of our apple trees, so I decided to give it another chance. This time, I overlooked the massive vine and slow road to flowering and maturity and focused on the fruit. Full of seeds and looking uglier than a spiny stone fish if I shut my eyes, it really is good to eat!

Soil: It will grow in poor soil conditions but prefers good, fertile soil.

Position: Needs its own growing space. Preferably on a trellis and with plenty of room.

Water: Does well in hot and humid conditions and dry spells but regular watering is best.

Top tips: Unlike a normal cucumber this vine grows and grows and then slowly begins to flower producing fruit at the end of the season as the weather cools. It is expensive to buy at the shops.

Uses: Eat it off the plant by scooping out the seeds and pulp with a spoon. Use in salads or salsas and in drinks.

Nutritional power status: The inside part of the fruit has a high moisture content, which can help you stay hydrated, with a good source of vitamin C and phenols, which help the body to eliminate toxins.

19. Fig

Easy – Fruiting can be temperamental

Why?: I know figs aren't that unusual and in many European countries they grow everywhere but you don't see them much in backyards in Australia. Figs are a delicious fruit and even better when preserved through drying as it intensifies the flavour.

Soil: Prefers lighter soils, sandy rather than heavy.

Position: Likes full sun where possible.

Water: Regular watering especially during the fruiting season.

Top tips: Cover the tree with netting the minute fruit starts to appear to deter birds and other pests. Fruiting might be impeded in hot and humid climates as figs tend to like warm, dry areas. Some farmers have had success in subtropical climates by growing under protection like greenhouses where they can better control pests and the temperature.

Uses: Freshly eaten from the tree, added to breakfast cereals, dried for a snack or BBQ, or enjoy with a dollop of vanilla yoghurt or ice cream.

Nutritional power status: This little dried fruit has fabulous laxative properties that can help boost a sluggish bowel. Also rich is calcium for strong healthy bones.

Ficus carica
Common Fig, Brown Turkey Fig

Glycine max (harvested young)
Young soybeans

20. Edamame

Medium — Fresh soya beans

Why?: Edamame isn't a crop that I would have naturally thought of growing if it weren't for our family deep-diving into Japanese food after several sushi bar-style restaurants opened in our area. We've become very fond of Japanese cuisine including the edamame appetiser, so while growing rice wouldn't be the best use of our garden space, growing edamame soya beans turned out to be a nice touch of Japan.

Soil: Well-drained, rich soil; doesn't like boggy soil or wet feet.

Position: Prefers full sun.

Water: Regular watering especially during fruiting period.

Top tips: Pick at the point of ripeness and don't let them dry out. Steam or boil and lightly salt for an instant snack. If they become overripe, they will be stringy and tough.

Uses: Cooked and salted makes a lovely appetiser or nibbles with drinks. Great vegetarian protein alternative.

Nutritional power status: Edamame are naturally gluten-free and low in calories. They contain no cholesterol and contain protein, iron and calcium.

20 crops I grew so you don't have to!

You would have realised by now I'm one of those food gardeners who just has to give everything a go even if the odds are stacked against me. I still grow several olive trees in the hope that one year they'll produce enough fruit to fill a pint jar for brining!

The point is, you never know until you try and there is a tonne to be learnt through experimentation. It's also fun and satisfying when you stumble on a fix that changes the game. Having said that, this final list of 20 food crops are hard for me to recommend because even though I still grow some of them they frustrate me no end.

One person's trash is another person's treasure, so there might be some things on this list that you totally disagree with me about and that's fine. But my reasoning why I grew them so you don't have to might interest you.

Here we go . . .

Brassica oleracea (Gemmifera Group)

Correct plural is actually Brussels sprouts

1. Brussels sprouts

Hard — If you live in a hot humid climate

Why?: I hate the taste of them and in our subtropical climate the growing season for Brussels sprouts is simply not long enough. The little cabbage-like things don't get a chance to develop before the weather heats up again and the plant perishes.

Soil: Free-draining, nutrient-rich soil with added nitrogen.

Position: Prefers full sun.

Water: Regular watering especially during fruiting period.

Top tips: Prefers a cooler climate and has an extra-long growing window before maturity. Watch out for cabbage butterfly as they will destroy the sprouts.

Uses: A leafy green addition to any roast meal, or fried with soy sauce and bacon as a stand-alone side dish.

Nutritional power status: From the brassica family, they're an excellent fibre source and contain zeaxanthin which is an antioxidant that may help promote eye health.

2. Starfruit

Easy – Kind of a shame

Why?: We have two starfruit trees and both fruit profusely! Unfortunately, shortly after planting them I found out about the warnings: too much of the juice from a starfruit can damage your kidneys. Not only that, substances in starfruit can affect the brain and cause neurological disorders. People with healthy, normal kidneys can process and pass this toxin from their body. However, for those with kidney disease, it's not possible. So yeah, it puts me off eating the fruit and I certainly wouldn't give them to anyone. In the future, I intend to remove these trees to make room for safer ones.

Soil: Well-drained, rich soil, add some compost.

Position: Prefers full sun where possible.

Water: Regular normal watering.

Top tips: Susceptible to fruit fly – pick the fruit as it ripens and don't let it spend time on the tree once ripe as they will get stung.

Uses: Adds sweetness to dishes; mix the juice with vegetables.

Nutritional power status: Starfruits are high in nutrients like vitamin C and copper. They also contain several antioxidants, including proanthocyanins and gallic acid. And they're a good source of insoluble fibre, which promotes digestive health and helps you feel full for longer.

Averrhoa carambola
Carambola

Diospyros digyna
Black sapote

3. Chocolate pudding fruit

Easy – They lied to me

Why?: I am not a fan of this fruit, it doesn't taste at all like chocolate pudding so I'm not sure why it's called that. Black sapote is its proper name and more appropriate than anything to do with chocolate.

Soil: Grows in most soil types and prefers subtropical and tropic climates.

Position: Likes full sun.

Water: Regular normal watering.

Top tips: Best to eat the fruit when overripe.

Uses: The texture is unique (like a custard or pudding), and can be added to cakes and baking for flavour like you would cocoa or cacao. Useful if you want to use something other than flour for baking cakes.

Nutritional power status: Low in fat and contains about four times as much vitamin C as your average orange.

4. Malabar spinach (climbing spinach)

Easy — Likes hot climates

Why?: I'm not a fan of the texture as it feels slimy in the mouth. It's like trying to eat a piece of aloe vera.

Soil: Well-drained soil with plenty of organic matter.

Position: Prefers full sun.

Water: Regular watering.

Top tips: It's a climber so best grown on a trellis. Store leaves short term in a perforated plastic bag in the fridge. For longer-term storage leaves can be blanched and then frozen.

Uses: Although not a true spinach you can use it like spinach or silver beet.

Nutritional power status: A good source of vitamin A and contains three times the amount of vitamin C compared to regular spinach.

Basella alba
Climbing spinach

Abelmoschus esculentus
Lady's finger

5. Okra

Easy – In warm climates

Why?: Initially I had high hopes for okra because I know many of my supporters and friends in the US love it. Problem is I can't get over the unpleasant texture in the mouth and the pods ooze a slimy substance that I don't like.

Soil: Well-drained soil with plenty of organic matter – animal manure or compost.

Position: Full sun – at least 6 hours per day.

Water: Water deeply in the early morning or late afternoon. Avoid watering the leaves to avoid fungal diseases.

Top tips: Okra pods are ready to harvest when they are 8–10 centimetres (2 inches) long. If the pods are left to grow larger, they will become tough and stringy. Grow a set of several plants so a quantity of pods can be picked at once.

Uses: Grown as a popular side dish vegetable, they're crisp and juicy with a dense and creamy texture. The flowers are also edible, making for an accompaniment to summer salads and drinks.

Nutritional power status: The vitamin C in okra helps support healthy immune function. It's also rich in vitamin K, which helps your body clot blood.

6.   Jicama

Easy – But poisonous

Why?: The beans are poisonous so it's risky to grow if you have pets or kids. You can only eat the tuber part and while that is not too bad it's not the best tuber as it's watery and lacks flavour. There are other better tuber veggies like potato, sweet potato and yams.

Soil: Prefers loamy, well-drained soils with plenty of nutrients and ample moisture.

Position: Full sun.

Water: Regular watering.

Top tips: Grow it up a strong trellis as it can get big. It needs a long growing season of at least 6–8 months for the single tuber to develop at the base of the plant.

Uses: Though most often eaten raw, such as chopped into salads, jicama can be steamed, boiled, sautéed or fried.

Nutritional power status: A root vegetable like potato or sweet potato with a much lower carbohydrate. The skin, unlike potato, is not edible.

Pachyrhizus erosus

Mexican potato, Ahipa, Saa got, Chinese potato, Sweet turnip

Cajanus cajan
Toor dal, Red gram, Gungo pea, No-eye pea

7. Pigeon pea

Medium – Very hard to shell

Why?: The process of harvesting each pea pod and removing the peas is laborious and takes way too much time. I paid my nephew to peel pods for one whole day and when he'd finished he only had about a kilogram. The foliage is supposed to make excellent mulch, but I don't think it's any better than other shredded plants or trees. To grow this crop properly you'd need to grow an acre or more.

Soil: A very hardy plant that grows in most soils.

Position: Full sun.

Water: Normal watering, is very drought tolerant.

Top tips: Can use as a living mulch, cutting back will encourage growth. Provides nitrogen which is good for other plants. May need an inoculant (bacteria) to germinate if not present in your soil natively. The peas last a long time in storage.

Uses: Soak and use as you would lentils, in soups and dips. Great for animal fodder and chicken feed if you can harvest enough of it.

Nutritional power status: Like other legumes in their nutrient content. A cup of uncooked pigeon peas provides 9 grams of protein and 9.5 grams of fibre which is essential for a healthy gut biome.

8.   Bitter gourd

Easy – Bitter truth

Why?: I've given this vegetable so many chances but they are just too bitter for my taste, even after I've salted and washed them. I know they're popular in Asian countries, so it pains me to not recommend them as they grow so well.

Soil: Well-drained, compost-rich soil.

Position: Full sun with at least 6–8 hours per day. Needs a strong trellis or gourd tunnel.

Water: Keep up the water for big and juicy fruits.

Top tips: Growing on a trellis will reduce risk of disease. Prune the tips from the upper branches to produce more low-lying fruit. Although the immature fruit is bitter, the ripe fruit has a pleasant-tasting pulp around the seeds.

Uses: An acquired, bitter taste, but can add a great flavour to relish and salsa. Can be stuffed with meat or seafood and baked.

Nutritional power status: High in fibre, beta carotene (a type of pre-vitamin A) and potassium, an electrolyte that balances fluid in your body.

Momordica charantia
Bitter melon

Fragaria vesca
Wild strawberry

9. Alpine strawberries

Easy – Just so irritatingly small

Why?: You can get larger alpine strawberries but most varieties are just too small and you're better off growing the regular-sized varieties. Some berries are so small that they shrivel on the plant before picking.

Soil: Well-drained soil, rich in organic matter. Prefers slightly acidic (pH 5.5–7).

Position: Prefers full sun but does okay in part-shade, and in containers.

Water: Water frequently when conditions are hot and dry as they can stress easily due to their shallow rooting system.

Top tips: Use an organic mulch to reduce moisture loss and protect the shallow roots. Growing from seed is a long and tedious process so best purchased as a seedling.

Uses: Smaller and sweeter than regular strawberries. They won't ripen further once picked so be patient and pick when ready to eat. Use in fruit salads, smoothies or straight off the vine.

Nutritional power status: High in vitamin C and B-group vitamins which help unlock the energy from your food.

10. Cucamelon

Easy – cute, mini-looking watermelon

Why?: Yes, they look cute but you may as well grow regular cucumbers. Their skin is tough and unpleasant compared to regular cucumbers . . . so why bother? I've tried pickling them and using them in various ways, but I always ask myself, 'Are they better than regular cucumbers? Are they delicious on their own?' Nope!

Soil: Rich, well-drained soil; very drought resistant.

Position: Full sun and prefers hot weather – a low-maintenance crop.

Water: Water thoroughly once a week and twice a week in hot weather.

Top tips: To regrow, pick a fruit from the vine and leave for around 1–2 weeks then open and remove seeds, laying on paper to dry out. Grow on a trellis for better control. Pick the little melons when young or the skins get tough.

Uses: Despite the look it is not a watermelon or a hybrid of watermelon and cucumber but it tastes like a sour cucumber and is edible, including the skin. Add to salads and salsa or tzatziki for use with a curry.

Nutritional power status: Low in calories and packs a nutritional punch, full of vitamins, minerals and fibre . . . if you can stand the taste!

Melothria scabra
**Mouse melon,
Mexican sour gherkin**

Helianthus annuus
Seeds and petals edible

11. Sunflowers

Easy – A controversial pick

Why?: I know this is a controversial pick because we all love big beautiful sunflowers but in my part of the world they attract cockatoos, which are lovely birds but become pests and are destructive in other parts of the garden.

Soil: Fertile, well-drained soil with slightly acidic pH. They are very hardy flowers.

Position: Full sun with at least 6–8 hours a day.

Water: Regular watering at the base of the plant, not over the leaves.

Top tips: Sunflowers might affect the growth of other plants by shading them out and via chemicals omitted to protect their space.

Uses: You can eat the seed straight from the sunflower, however roasting them at 150°C (300°F) for 5 minutes enhances the flavour. They attract bees to help pollinate other vegetables, salads and fruits.

Nutritional power status: Rich in protein, monounsaturated fats and B-group vitamins. These healthy fats are linked to lower rates of cardiovascular disease.

12. Brukale

Hard – Just like Brussels

Why?: I don't like Brussels sprouts but I do like kale and since this vegetable is a cross between Brussels sprouts and kale I stupidly thought I might like it. It's all academic anyway because just like Brussels sprouts this variety is also too hard to grow in our subtropical climate. They need a longer season of cool weather to flourish.

Soil: Fertile, free-draining soil.

Position: Sunny and sheltered from winds.

Water: Water often in warm weather.

Top tips: Grow in a seed tray first to germinate or directly in a garden bed. Give them plenty of room as they can grow up to a metre (3 foot) tall and have large leaves.

Uses: The entire plant is edible from the tender base to the leafy top. You can roast, steam, stir-fry or grill the leaves and use them to wrap food such as minced meat or chicken breast. Also a fresh addition when added raw to salads.

Nutritional power status: High in vitamin K which helps with blood clotting.

Brassica oleracea hybrid
Cross between Brussels sprouts and kale

Brassica juncea
Japanese giant red mustard

13. Giant red mustard

Easy – Looks beautiful in the garden

Why?: They are very pungent and not everyone's cup of tea. I'd rather grow other mustards that have better-tasting leaves, stems and seeds. Having said that, giant (or Japanese) mustard does look amazing in the garden and its flowers attract bees and good insects from miles away.

Soil: Hardy and grows in most soil types, free-draining and fertile. Can add some lime to bring the pH to slightly above 7.

Position: Full sun but will tolerate part-shade.

Water: Regularly but is quite drought hardy. You'll get a better plant and a crisper leaf if you water it more.

Top tips: Pick early if you don't like too much of a zing. It gets hotter with age so leave on the plant if you want more spice in your life! It regularly goes to seed and re-sows itself. Quite self-sufficient in its own right!

Uses: Harvest the seeds and use to spice up curries, use leaves in salads to add some heat or add some heat to your stir-fries.

Nutritional power status: Like most 'green leafy' vegetables, they're high in vitamin C and vitamin K.

14. Chinese potato

Easy – Small tuber plant to grow

Why?: They are just *too* small. Very finicky to harvest and a lot of work to peel and cook. I'd rather grow regular potatoes as they taste better and are bigger and easier to work with.

Soil: Likes the same soil as regular potatoes: well-drained and crumbly.

Position: Full sun for best results.

Water: Loves lots of water, especially during hot summers.

Top tips: Harvest all the potatoes after the season otherwise the ones you leave behind will rot in the soil.

Uses: As you would normal potatoes. Boiled, baked, roasted, barbequed, mashed or added as a bulking ingredient to fish caked, curries, soups and stews. Chinese potatoes are sweeter than regular potatoes.

Nutritional power status: Rich in potassium, the electrolyte that balances with sodium to maintain fluid balance in your body to keep your heart beating.

Plectranthus rotundifolius
Native potato (Africa)

Capsicum chinense
**Includes Carolina Reaper,
Bhut Jolokia**

15. Scorpion or ghost chillies/peppers

Easy – But why?

Why?: They are just too hot . . . Unless you're a weapons-grade scientist, they'll blow your skull off! Plus, these fruits can be dangerous. I never grow chillies just for the heat. The most important thing for me is the flavour. I prefer hot chillies such as habanero, jalapeño, Thai and African birdseye because they also have a nice tangy flavour.

Soil: Normal, fertile, free-draining soil.

Position: Full sun and loves hot weather.

Water: Keep well watered, especially in hot climates.

Top tips: Prune right back at the end of the season, over the winter months. It will appear dead. Give it good mulch and fertilise in spring. The second season can sometime result in better harvests.

Uses: Adds a super-hot hit to any meal! Not for the faint-hearted though.

Nutritional power status: Has an insanely high level of capsaicin which is touted as having benefits including improving heart health and aiding metabolism.

16. Perennial capsicum

Easy – Good producer

Why?: This plant is a genuine disappointment for me because I've tried to love it but I just can't get there. The skin on this variety of sweet pepper is very thin and doesn't taste great so I'd rather grow a sweet pepper, larger capsicum or a tasty chilli.

Soil: Well-drained, fertile soil.

Position: Full sun.

Water: Water when necessary but don't let the plant dry out otherwise it will only bear small fruit.

Top tips: Stake the plant as it can get tall and topple over under the weight of the fruit.

Uses: Chop straight into salads, stir-fries, stuff with beans and rice or mince dishes. Grill or roast and store in oil for an antipasto.

Nutritional power status: Rich in phytochemicals that may have anti-inflammatory and antioxidant properties.

Capsicum pubescens
Rocoto pepper

Musa spp.

Includes Cavendish, Lady Finger, Red Dacca

17. Tall banana plants

Easy – I'm too old to climb trees

Why?: Bananas could easily have been in my top 20 fruits if it wasn't for the problem of climbing them to bag the bunch, which is what you have to do to protect them from animals. They can grow up to 5 metres (15 feet) tall and are simply too dangerous to bag climbing up high ladders. They also require a lot of space to grow. I'm growing dwarf and super dwarf varieties which are doing well and showing promise.

Soil: Most types of soil but they just hate boggy, non-draining soils.

Position: Full sun.

Water: They love lots of water but make sure the ground is free-draining – they hate 'wet feet' and their roots will rot.

Top tips: Use adequate fertiliser regularly as they are hungry plants that produce lots of fruit. You will get a bigger bunch if you feed them well. Cover the growing bunch if you have an animal or bird issue.

Uses: Obviously eaten straight off the tree. Use in fruit salads, cooking and baking, use less ripe (green bananas) in curries and savoury dishes.

Nutritional power status: Rich in vitamin C, B-group vitamins, fibre and potassium. I don't know many people who don't like bananas!

18. Broad beans

Medium – I'll stick with climbing beans

Why?: The Brussels sprouts of beans. I don't really like the taste of broad beans and can't see the value in growing them so I would rather grow something else instead. They also need support and are easily blown over and damaged in windy weather.

Soil: Fertile, free-draining soil.

Position: Full sun.

Water: Make sure they have adequate water and don't dry out.

Top tips: Use a trellis as this type of plant quite commonly falls over.

Uses: A great protein alternative to meat. Use in soups, casseroles, curries and vegetarian patties.

Nutritional power status: They are rich in protein and fibre and contain zinc and iron so are an excellent substitute for meat for a nourishing vegetarian dish.

Vicia faba
Fava beans

Cicer arietinum
Garbanzo beans

19. Chickpeas

Medium – High maintenance

Why?: I originally thought of growing chickpeas to make organic hummus but then I found out they take up a lot of your garden for a small return. They are fiddly to harvest so not a great use of time and space and they're susceptible to pest damage in warmer climates. This is one crop I'm happy to buy in a can.

Soil: Fertile, free-draining soil.

Position: Full sun.

Water: Water regularly and don't let dry them out.

Top tips: They are better as a farm crop as they need plenty of room.

Uses: If you do grow them, they are a tasty alternative to animal protein. Use mixed into a dip, in salads, in fritters or vegetarian meat balls. The nutty, grainy texture pairs well with many vegetables and salads. Used extensively in Middle Eastern dishes.

Nutritional power status: Rich in iron and zinc for a healthy immune system, and phosphorous and magnesium for your bones and muscles.

20. Turkey berry

Easy – But no thanks

Why?: My interest in growing turkey berry (or pea eggplant) began on a tour of Thailand back in 2016. Nina and I took a cooking class and made Pad Thai with these small green berries. Turkey berry can become unruly in your garden and grow to 3 metres (10 feet) high so be careful where you plant it. The berries are bitter but with salting and cooking they do improve. To be honest, I reckon I'm better off growing the larger, modern style of eggplant! The turkey berry was crossed with other varieties to produce today's eggplants that are bigger and less bitter so the good thing about growing this plant at least once is knowing that you've produced something old and original that hasn't changed for thousands of years.

Soil: Grows in a wide range of soils from dry to boggy. A hardy plant.

Position: Full sun to part-shade.

Water: Likes a lot of water, especially as the fruit has a high water content.

Top tips: It has large spines that grow on the stems that really hurt if they stab you! It can become a weed so be careful with its spread through your garden; contain it if you can.

Uses: A key ingredient in Pad Thai for the purists but I don't miss it personally.

Nutritional power status: Low in energy and carbs it can be used instead of starchy vegetables to help control blood sugar and weight gain. Contains antioxidants for general health and fibre for bowel health.

Solanum torvum
Pea eggplant

My top 100 on reflection

Looking back through these five lists of my top 20 food crops, I've been struck by how common most of my selections are . . . I guess I like to grow what my family and I enjoy eating, which isn't too different to most people or, dare I say it, what's commonly available at the supermarket!

At the same time, some of the fruits, veggies and herbs *are* a little more unusual and one of the main objectives of this book is to shed some light on them and encourage you to try growing them yourself. No doubt there are some plants on the lists that will be tricky or controversial, but I hope that's balanced by others that you'll see as gems you're inspired to try. So what are you waiting for?

Let's get gardening!

The joys of gardening

It may sound cliché, but gardening truly brings joy to both the heart and mind.

Over the past decade, I've received hundreds of thousands of emails and messages — and among them, dozens have shared how gardening literally saved their life. It pulled them back from the brink. . .

So, why is gardening so powerful?

I think because connecting with nature isn't just important — it's essential. Nature is where we come from and it's where we belong.

When you combine that connection with the satisfaction of nurturing plants, watching them grow, and creating a space where birds, insects, and wildlife feel welcome, the result is something profound.

And then, of course, there's the tangible reward: harvesting the fruits of your labor.

That combination — purpose, peace, connection, and abundance — multiplies the joy in ways that are hard to explain . . . but easy to feel.

Preserving Your Harvest

Often beginner backyard food growers worry that they won't grow anything. The truth is they should be more concerned about what they will do when they grow more food than they can eat! This is a real and great problem to have!

Obviously, it's always nice to give excess food you have grown with love away to friends and family, but you can also preserve a glut of produce to eat later. Plus, preserving food can give you an opportunity to enhance the flavour or make them even healthier!

I won't delve too much into how to do these methods, but essentially, there are six main ways I like to preserve food, and they are: Freezing, Canning, Pickling and Jams, Dehydration, Fermentation, Freeze Drying.

Sauerkraut (cabbage)

Apple cider vinegar

Tomatoes

Zucchini

Fermented Dill Pickles

Lacto-fermented perennial
capsicum and cucumbers

Fermentation

Sounds like weird science but once I realized that so much good food we eat and drink are made through fermentation it's not that scary. Think of coffee, beer, cheese, yogurt, chocolate, vinegar, sourdough bread, and you'll get over the yuk factor quick smart. Sauerkraut, kimchi, and dill pickled cucumbers are the most famous fermented veggies, however, with a bit of salt mixed in spring water you can ferment any vegetable and turn it into a tasty, preserved snack.

Dehydration

Is simply the process of drying food to get the moisture out of it because without moisture bad bacteria can't grow and thus spoil your food. Dehydration can be used in combination with other forms of preserving, for instance, salting tomato pieces before sun drying them or using a dehydrator or even your oven to dry the tomato and turn them into a tasty, preserved snack. In this case, olive oil can be used acting as an extra preserving agent and to improve the taste of the dehydrated tomatoes over time.

Mango

Lime slices

Yes, we have some bananas

Pickled tumeric

Mustard stem

Pickling and Jams

Archaeological evidence suggests that people in northern China were using salt to preserve food more than 8000 years ago. It's true! Pickling with vinegar came a bit later at around 5000 years ago in the Middle East and preserving with sugar is only a recent thing starting in India about 1500 years ago. There's no doubt making jams and pickling cucumbers is in your blood so give it a go!

Rosella jam

Egglant preserve

Freeze Drying

Freeze dried food, incredibly, was first used by the Incas centuries ago by using natural freezing and thawing to remove moisture from food so it could be stored. Modern day freeze drying began in the 1930s using big industrial machines. Today, you can buy household appliances the size of an oven and freeze dry food at home – it's a game changer! Freeze dried vegetables can be kept and stored for 25 years. I'm not sure why you'd want to store your homegrown produce for that long, maybe you want to shock the grandkids, nevertheless, it is impressive.

Freezing

This is one of the fastest and easiest ways to preserve food. Most frozen veggies and fruit will keep for 6-12 months.

For example, too many potatoes can be cut into chips, blanched, and placed spread out on cooking wax paper in a tray in the freezer to freeze individually. Once frozen, pack into a container or freezer bags to fry up later at your leisure.

Canning

Just like canned corn, peas, or baked beans from the shop, you can do the same thing with your veggies to preserve them and consume later. Home canning requires some technical knowhow, but it isn't hard. Most people use glass jars to can their food and one big advantage of canning is the ability to store them at room temperature without needing refrigeration.

and some more ideas for inspiration!

Dried rosemary

Dill powder

Tumeric four ways

Horned cucumber salsa

Pigeon pea dip

Chilli sauce

**Salted limes/lemons
(Moroccan style)**

Tamarillo fruit balls

Fruit leather

And there's nothing better than fresh out of the garden!

Keep up with me at:
selfsufficientme.com